Broadband Communications and Home Networking

Broadband Communications and Home Networking

Scott R. Bullock

NOBLE
PUBLISHING

Noble Publishing Corporation
Atlanta, GA

Library of Congress Cataloging-in-Publication Data

Bullock, Scott R., 1950-
    Broadband communications and home networking / Scott R. Bullock.
        p.cm.
    Includes bibliographic references and index.
    ISBN 1-884932-19-3
        1. Home computer networks. 2. Broadband communication systems. I. Title.

TK5105.75 .B85 2001
004.6'8--dc21   2001032692

N O B L E
PUBLISHING

Printed in the Unites States of America
ISBN 1-884932-19-3

To my loving wife, Debi;
to Crystal, Cindy, Brian, Andy, and Jenny,
and to my mother, Elaine

Contents

The "Last Mile"   109

Satellite Communications   121

Index   143

Introduction

Broadband communications are high-speed digital communications that require a wide bandwidth for transmission. Commonly, broadband communications provide a technology that brings and distributes high-speed data, voice, and video to and throughout the home. It is used for networking, Internet access and distribution and provides a means of connecting to the outside world without installing new wires.

Information is brought to the home using different methods, such as phone lines, wireless RF, fiber optic cable, coaxial cable, and satellite links. The information generally comes into the home at a single location. Therefore, a means for distributing this information throughout the home is required. Along with the distribution of information, networking plays an important role for the connection and interaction between different devices in the home.

How is the information delivered to the end user? There are five methods of bringing information to the home: telephone wire, cable, fiber optics, wireless and RF, and satellite communications. Three types of distribution mediums are used to distribute high-speed voice, data, and video throughout the home: transmission over the power lines, phone lines, or through the air using RF communications.

Broadband communications and home networking are becoming household words as more and more homes are utilizing multiple PCs. As the technology curve climbs to new heights, household appliances, security systems, lighting, environmental control, televisions, stereos, and nearly anything that is electrical in the home is becoming digitally controlled and adaptable to a network system. To provide an ubiquitous solution to home networking worldwide, an easy-to-install plug-n-play system needs to be available at a reasonable cost to end users.

Basic Principles of Telephony

Many of the basic principles of the original telephone operation and nomenclature are still in existence today. Broadband technology and home networking brings about radical changes in the existing infrastructure and operation of the standard telephone and the various types of signals that will be delivered along with voice, including high-speed data for use with the Internet and other applications, music, video including *high-definition television* (HDTV) and other new applications. As the ways and methods that information is delivered to the subscriber is changing rapidly, enhancements to the standard *public switched telephone network* (PSTN) and other means to connect the subscribers, including wired and wireless solutions are being put into place.

This chapter reviews the basic principles of telephony and points of interest to aid a basic understanding of how the telephone operates and how the emerging new technologies affect telephony today. Discussed are telephony circuitry, the basic infrastructure, and alternatives to running the signals to the home.

1.1 History

The telephone was invented by Alexander Graham Bell in 1876. For several years, American Telephone and Telegraph (AT&T) owned nearly all of the telephone services. The telephony research was conducted by Bell Labs. The AT&T monopoly was eventually broken up and *regional Bell operating companies* (RBOCs) were created. The *central service organiza-*

tion (CSO) became responsible for developing new technologies and perform the research and development for all of the RBOCs. Bell Labs continued their research for AT&T and was later renamed Lucent Technologies. Currently. Since AT&T lost control of the local distribution of telephone signals and the RBOCs are not allowing AT&T to use the existing telephone infrastructure, the company approached the "last mile" to the home by purchasing TCI Cable Company, which is now called AT&T cable services. With the infrastructure of cable in place, AT&T is combining cable services with telephone and data services through the cable instead of the telephone wire and telephone company. This has helped to fuel the different ways of bringing high-speed information to the home.

The *central office* (CO) handles the local switching, i.e., it connects the two communicating telephones together along with other type of telephony functions and features, including *caller identification* (CID), call waiting, and other on-hook and off-hook services. With the demands for connecting to the Internet and data modems, the existing telephone line infrastructure is used to send digital data information to and from the subscriber. As the data rates increase, other methods are being utilized to accommodate the increased speeds and bandwidths that are required by the end user. New wiring, including fiber optics and coaxial cable are used to bring higher data rates to the user as well as ensure higher quality and reliability. Satellite communications is another means of bringing high-speed data to the subscriber and provides the bandwidth and wide range of coverage necessary to keep up with the demands of the end user. As these technologies are developed and installed, the present telephone system is going to be changing at a rapid pace. Exactly what method or methods will be used in the future for providing high-speed information transfer between the subscriber and the provider is still to be determined.

1.2   Telephony Fundamentals

A typical telephone wire, the PSTN or *plain old telephone service* (POTS), is connected to the CO. The telephone wire consists of four colored wires that are connected to a standard telephone jack, RJ-11 (see Figure 1-1). In most connections, the inner pair of wires, green #2 and red #3, is used for the *tip* and *ring,* respectively, for a basic single line telephone. The inner pair of wires is used to connect to a standard telephone or telephone device to complete the local loop to the CO. The outer pair of wires, black #4 and yellow #1, is used for the *tip* and *ring,* respectively, and provides another line connection to the CO. The outer pair is often used for two-line phones or for an additional connection to the CO that could be used for data applications. This additional pair completes another local loop between the CO and the subscriber (see Figure 1-1).

Plugs Into Telephone
Or Telephone Device

3-Red
"Ring1"

2-Green
"Tip1"

4-Black
"Tip 2"

1-Yellow
"Ring 2"

RJ-11
Telephone Plug

Telephone
Wire

To Central Office or
Wall Connection

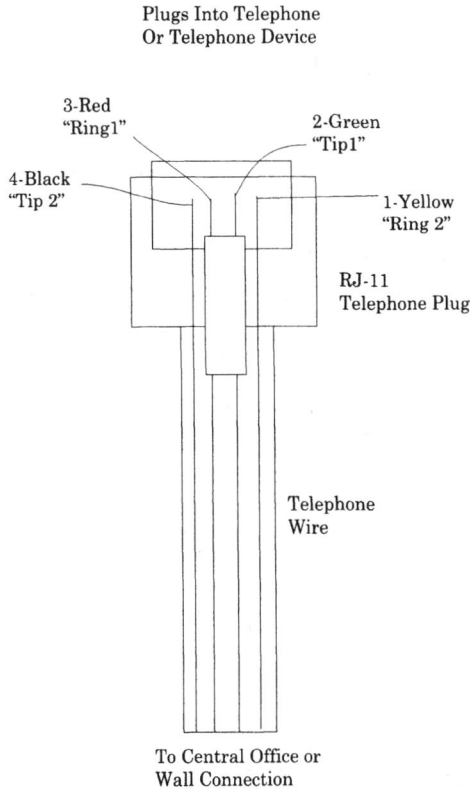

Figure 1-1 *Telephone wiring in the home.*

The Local Loop

The local loop connects the central office to the home or subscriber. This connection uses two wires; it is called a loop because it comes from the central office down the tip to the subscriber and back to the CO on the ring wire, thus forming a loop (see Figure 1-2).

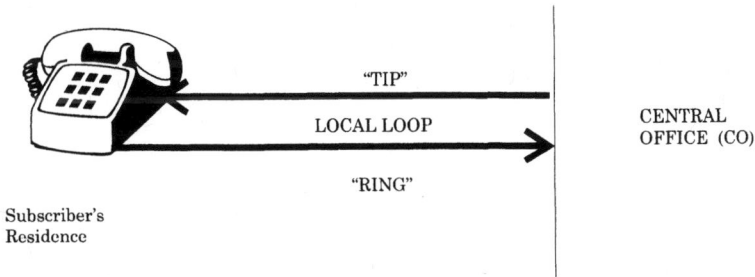

"TIP"

LOCAL LOOP

CENTRAL
OFFICE (CO)

"RING"

Subscriber's
Residence

Figure 1-2 *The local loop.*

The loop completes the connection and provides information to and from the subscriber. Along with the voice or data that is sent and received, the loop also provides all of the control and detection information, including ringing signals to ring the phone, detection of the line to see if the line is available or in use, detection of a telephone that is left off hook, the number that is dialed so that the CO can connect to the desired party, and many other types of communications and features.

The terms *tip* and *ring* come from the old days of manually connecting the telephones together. The *tip* was the tip of the plug and the *ring* was the upper part of the plug (see Figure 1-3a). When someone would ring the operator, the operator would manually plug in the wires from the caller's connection to the called connection. The plug would contain both the *tip* and *ring* for connecting the local loop. These terms continue to be used to describe the connections between users for the local loop.

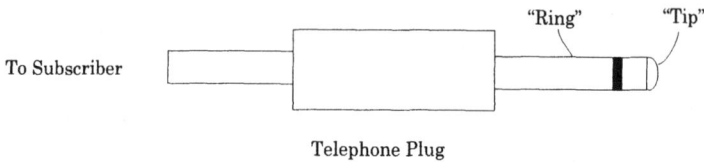

a. Plug used to connect users showing "tip" and "ring".

b. Manual telephone that relied on a live operator to make the connection.

Figure 1-3 *a. Telephone plug used by the operator to connect calls. b. Manual telephone that relied on a live operator to make the connection.*

The first telephones used a live operator to connect the telephone lines. A typical telephone is shown in Figure 1-3b. The person making the call would lift up the transmitter of the telephone and talk to the live operator. The person would tell the operator whom they would like to talk to, and the operator would then ring the other person's phone to let them know there was a call. If the person answered the telephone, then the operator would connect the two parties together. The microphone and the earpiece were generally separate units, with the earpiece on a cord so that it could be brought to the person's ear for easier listening (see Figure 1-3b).

On and Off Hook

The CO provides a battery *direct current* (DC) voltage across the *tip* and *ring* wires equal to approximately +48 volts DC. The CO uses this voltage to determine if the phone is on or off hook. If the phone is on-hook (the receiver is on the cradle of the phone or hung up), then the switch hook is opened and no current is flowing. The CO measures a DC voltage of 48 volts, which informs the CO that the subscriber is on-hook or hung up. Except for any on-hook signaling that is requested, the CO remains dormant for that particular subscriber.

If the receiver is taken off hook (the receiver is lifted off of the cradle or turned on), a switch in the telephone is closed to allow current to flow and the DC voltage is lowered (see Figure 1-4). The CO senses this drop in voltage and determines that the subscriber's telephone is off hook. If this is done in answering a ring from the CO, then the CO knows that the connection is completed and the two parties are in communications with each other. If the subscriber is not connected to another party through the CO, then the CO will assume that the telephone was left off-hook and will send a audio signal to the subscriber's telephone informing them to hang up the phone.

Telephone Warning Signal
The CO detects a DC voltage when the subscriber is not connected to another subscriber. The CO then sends an off-hook warning signal which is a combination of four different frequencies. It is turned on for one-tenth of a second and shut off for one-thenth of a second and is sent at a very high signal level or volume so it can be easily detected by the user.

Telephone Busy Signal
If the CO is trying to connect a request from one subscriber and determines, by detecting the DC voltage drop, that the requested subscriber is off-hook, the CO sends a busy signal to the requester. The busy signal is a tone that is on for one-half second and off for one-half second.

Telephone Ringer and On-Hook Signaling

When the subscriber's telephone device is on-hook or hung up, the weight of the telephone receiver, when it is resting in the cradle, opens up the DC path and keeps the switch open. However, there is a capacitor that AC couples both the ringer circuitry and the on-hook signaling, including caller ID and other types of functions, while the telephone is on-hook. These signals are required to be received by the subscriber's telephone circuitry to allow them to be detected while the telephone is on-hook (see Figure 1-4).

The ringer circuit requires a high voltage DC in order to ring the telephone device. This requirement comes from the old days when telephones had a mechanical ringing device that struck the bells in the telephone and has carried over for today's requirements, so that the circuitry has the capability to work with older style telephones. The telephone circuitry inside the phone provides a high impedance path to the ringer signal so that the voice and data information signals are not affected by this high voltage ringer signal.

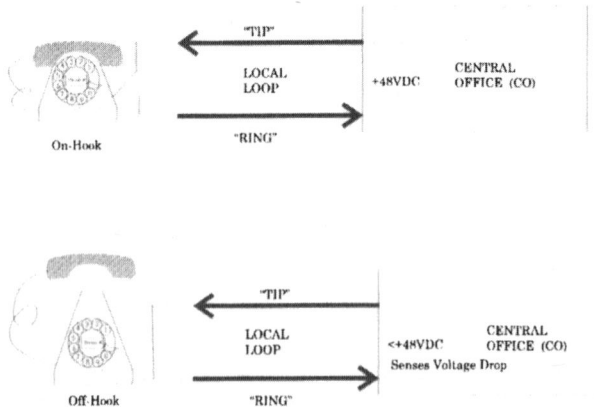

Figure 1-4 *On/off hook voltage detection of a standard telephone.*

Pulse Dial/Tone Dial

Pulse dial was used exclusively in the start up of telephony. It is nearly phased out and seldom used today. Both pulse dial and tone dial are methods to let the CO know what number the requesting subscriber is trying to reach. The CO detects the number by either counting the number of on/off pulses or by decoding the tones to numbers. Once the CO has decoded the pulses or tones into numbers, it knows to which subscriber to connect the request.

Pulse Dial

When a telephone is used with pulse dial, the dialer is usually manually turned and opens and closes a switch. The number of pulses caused by the switch determines the number that was dialed. Some of the old style phones had dials that required turning. When the dial was turned around, it opened and closed the switch (see Figure 1-5a). Therefore, if a "1" was dialed, the dial was turned only a very short distance so the CO would detect one pulse. If a "0" was dialed, the dial would turn its maximum distance; the switch would operate ten times and the CO would detect that a "0" was dialed.

Tone Dial

Tone dial has made pulse dialing nearly obsolete. Tone dial uses two different tones, between 350 and 440 hertz, which are sent out simultaneously for each number dialed, usually from a push button phone. Nearly all telephones use this *dual tone multiple frequency* (DTMF) system for sending numbers to the CO. An example of a push button tone dial telephone is shown in Figure 1-5b.

Many of the telephones came equipped with both DTMF and pulse dialing. The mode of operation was generally selected by manually moving the switch in the desired position. The pulse dial feature for these types of telephones would automatically open and close the switch, depending on the number dialed.

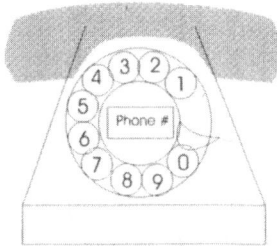

a. Rotary telephone using pulse dial.

Figure 1-5 *Pulse dial operation for a standard telephone.*

Sidetone

A sidetone represents a portion of the audio signal which is sent to the earpiece of a standard telephone so the person knows how loud to talk. If the sidetone is too soft, the person will talk too loud; if the sidetone is too loud, the person will talk too soft.

1.3   Basic Telephone Design

Connecting to the CO requires certain design and impedance characteristics. All devices that are connected to the CO have to pass the requirements specified in Part 68 of the FCC rules and regulations. These requirements help to keep the phone lines clean from noise and voltages and help to minimize reflections due to mismatches at the subscribers end.

Telephone Bandwidth

The maximum bandwidth associated with twisted pair copper wire is higher than that is required for voice or telephone applications. However, the bandwidth for the standard telephone line is limited by the PSTN to 4 kHz. The bandwidth required for voice type signals is from 0 to 4 kHz. For speech only, the bandwidth is limited from 300 hertz to 3 kHz. Sending data at high data rates using PSTN is difficult. New ways of connecting from the CO to the subscribers, including different methods of cabling and wireless techniques to achieve higher bandwidths capable of solving the high-speed data demands need to be found.

Impedance and DC Resistance

Since several types of devices, such as modems need to be connected to the PSTN line from the CO, there are basic design requirements for this type of connection. The general impedance for a basic connection to the telephone wires is 600 ohms. Other types of impedance matching circuits are used to provide the 600-ohm impedance, including resistor and capacitor combinations.

The DC resistance from the CO to the subscriber is specified and limited to 1500 ohms. This allows the use of this system with a range of approximately three miles using a wire gauge of somewhere in the order of 26 AWG.

RC Droop

For lines that are long, the signals experience degradation and amplitude variations the further down the line the subscriber is located. These variations are known as RC droop. RC droop describes the attenuation effects of the resistance and capacitance with regards to the frequency dependence of the amplitude response of the signals.

To mitigate these amplitude variations in the lines, lumped inductor coils with values at approximately 88 mHz are placed in the line, at approximately 6000 feet intervals, to provide a flat response. These inductors help to cancel out the effects of the resistor/capacitor RC rolloff, which helps to mitigate the RC droop. The bandwidth of the lines starts to roll off at 3 kHz. Problems with installing these inductors on the existing lines are cost and time efficiency and, with the desire to use these lines for data products, these inductors need to be removed for DSL services.

Coupling Voice to the Telephone Lines

Coupling voice signals to the telephone lines requires voice modulation. One of the methods that was used extensively in older phones is amplitude modulation of the voltage with an acoustic transducer. The microphone part of the telephone consisted of a diaphragm that moved back and forth from the acoustic waves generated by the speech at a frequency corresponding to the frequency content of the voice signal. As this diaphragm moved in and out, it compressed a module made of carbon (see Figure 1-6).

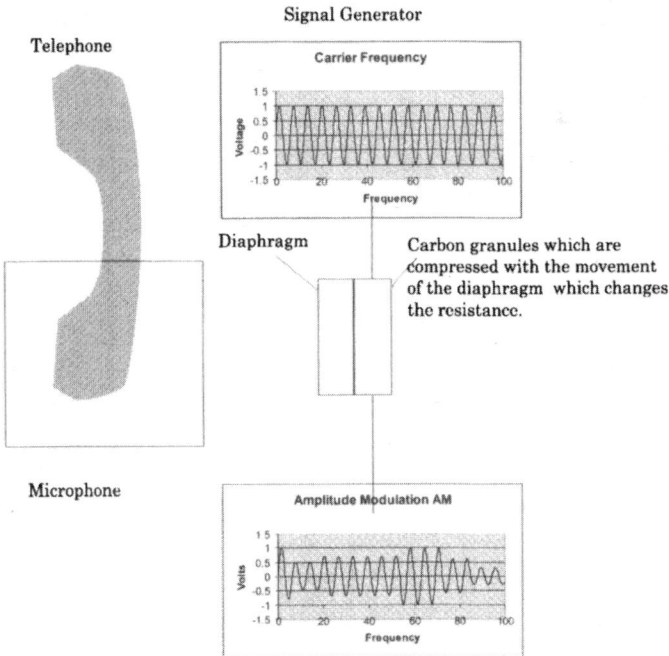

Figure 1-6 Basic operation of the transmitter portion of the telephone.

When the carbon granules are compressed, the resistance changes. Therefore, the current through the carbon changes with respect to the voice signal and at the frequency of the voice signal. The current through a resistor develops a voltage, and the variation of the current causes a variation in the amplitude of the voltage, known as *amplitude modulation* (AM).

The voice signal would amplitude-modulate the voltage. This signal would be connected to the other party via the CO, and the receiver on the other end would detect this amplitude modulation by a simple diode detector, amplify it if necessary, and feed this amplitude to the speaker on the receiving end (see Figure 1-7).

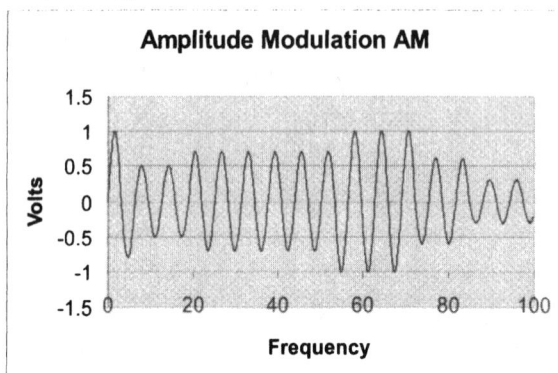

Figure 1-7 *Amplitude modulation voltage waveform.*

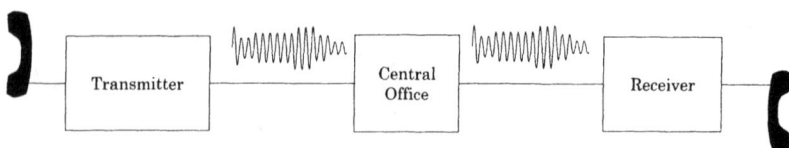

Figure 1-8 *Basic operation of the receiver portion of the telephone.*

Alternative methods need an acoustic-to-electrical transducer to convert acoustic waves into electrical signals to send the voice over the electrical line or over PSTN lines.

1.4   Three Types of Exchanges

There are three types of exchanges that handle the switching or connecting the telephones together: local, long haul, and exchange. The *local* is the central office and handles the switching or connecting of the local telephones. When a subscriber makes a call to another person in the same basic vicinity, then the local central office connects the requesting party with the called party and the connection is made. No other facility is involved with this connection. This local exchange is used for short-range connections.

The *long haul* is used to cover long distances. These connections are nearly all fiber optics to reduce the attenuation for the long lines. When the subscriber makes a long distance call, the long haul provides the connection over long distances.

The *exchange* is used as the intermediate connection between the local and the long haul. The subscriber makes a call and the local CO accepts the call, sends it off to the exchange, when it is handed off to the long haul. The call is routed through the exchange to the CO and to the called subscriber.

1.5   Modems

Modem is short for modulation/demodulation. Modulation is a technique used to transmit digital signals. Generally, digital signals are not sent directly from the transmitter to the receiver because a carrier is needed to transport the digital signals from one point to another. The digital signals modulate a carrier or *continuous wave* (CW) frequency signal in order to send it through a medium. The medium could be a wired line, which includes twisted pair or PSTN line, coaxial cable, and fiber optics, or it could be a wireless RF or microwave connection that is transmitted through the air. In either case, the digital signal modulates the carrier using various modulation schemes, including *phase-shift keying* (PSK), which changes the phase of the carrier frequency, *frequency shift keying* (FSK), which changes the frequency of the carrier, and *frequency hopping* (FH), which also changes the frequency to many different frequencies. There are several variations of the basic modulation schemes that will be discussed in later chapters.

Once the digital data is modulated onto a carrier frequency, the digital signal is "carried" by the carrier frequency to the receiver part of the system, which contains the demodulator. Figure 1-9 shows an example of a method for sending digital data on a carrier using *binary phase shift keying* (BPSK).

Another reason for using a carrier in wireless systems is that the antenna size for sending the digital signal directly would be impractically large. Therefore, by modulating the digital data onto a higher frequency, the size of the antenna required is much smaller and more practical.

The demodulator strips the carrier or eliminates the carrier and retrieves the digital data for processing in the receiver. Therefore, a modem performs both modulation/transmit and demodulation/receive.

Digital Data

BPSK Modulation

Retreived
Digital Data

Modulator

Demodulator

Carrier Frequency

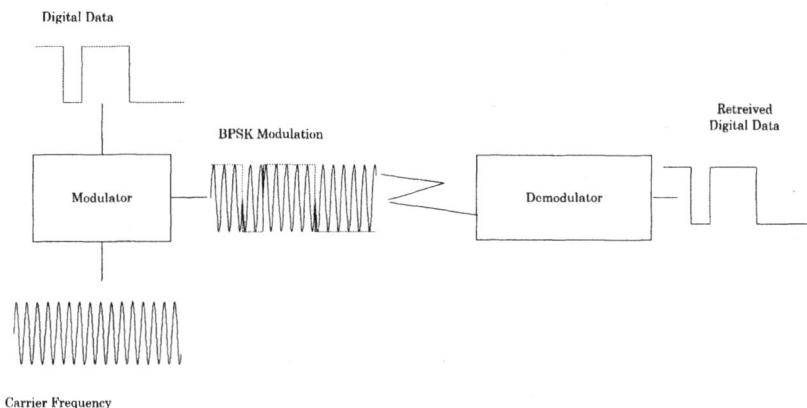

Figure 1-9 *Modulation/demodulation of digital data.*

Modems have gained tremendous popularity with the age of the Internet. Consumer interest in "surfing" the Internet has generated a very high demand on data throughput and speed to handle both the increase in traffic and also to download large files. There is an ever-increasing desire to increase the speed. New methods and standards are established as these speeds are increased to meet the demands. With the rapidly changing technology, standards are being set almost on a continuous basis. Some of the more popular standards established by the International Telecommunications Union (ITU) include the following:

V.22 – 1200 bps
V.22 bis – 2400 bps
V.32 – 9600 bps
V.32 bis – 14.4 kbps
V.34 – 28.8 kbps, 33.6 kbps
V.90 – 56 kbps*

*These modems are very sensitive to noise on the phone line and distance from the CO. Data rates for these modems range between 40 to 50 kbps, depending on line conditions.

The data rates achieved by these modems depend on the quality of the connection and wiring, the distance from the CO, and many other factors. The higher the data rate of the modem, the more these factors will affect the speeds.

Several different modulations schemes are used in modems. They include, but are not limited to, *quadrature phase shift keying* (QPSK), 8-phase shift keying (8PSK), *quadrature amplitude modulation* (QAM), and FSK. The more possible states there are in a given modulation scheme, the higher the data rates. This requires more sophisticated designs since these systems are more sensitive to noise and need a higher *signal-to-noise ratio* (SNR).

Other types of modems are currently being developed to provide higher speeds to the end user are *integrated services digital network* (ISDN), *digital subscriber loops* (xDSLs), and cable modems. The xDSLs include *asymmetrical digital subscribe loop* (ADSL) and *high-speed DSL* (HDSL).

1.6 T Carrier Systems

T1 carriers are used for transmissions of voice and data. The T1 carriers use *pulse coded modulation* (PCM) and *time division multiplexing* (TDM) in order to transmit digital signals to the subscriber. The T1 carriers are used to provide short-haul transmissions; each line is capable of handling 24 simultaneous voice-band signals. The T1 line can transmit and receive voice and data signal over a range of 5 to 50 miles. However, repeaters are required approximately every mile. Each of the repeaters receives and amplifies the incoming signals to give them a boost to travel over the next mile or so to the next repeater.

T2 carriers use the same modulation scheme as the T1 carriers, PCM/TDM. The T2 carriers are capable of providing 96 simultaneous voice-band channels using a single 6.312 Mbps data signal. The T2 carriers use a specially made cable assembly with very low capacitance. Since the capacitance is the main cause for rolloff and attenuation, this line can be used for transmission of up to 500 miles. The T2 carrier system provides higher data rates and is capable of carrying a single picture phone signal.

T3 carriers also use PCM/TDM to provide 672 voice-band channels. The T3 carrier system provides even higher data rates than the T1 or T2 lines reaching up to approximately 46 Mbps.

The T4M carriers also use PCM/TDM to provide 4032 voice-band channels and the T5 lines use PCM/TDM to provide 8064 voice-band channels with data speeds up to 560.16 Mbps over a single coax. A chart showing the different T carrier systems is presented in Table 1-1.

Table 1-1 *T Carrier Systems.*

System	Modulation/Multiple User	Voice-band Channels	Distance in miles	Data Rates
T1	PCM/TDM	24	50*	
T2	PCM/TDM	96	500**	6.312 Mbps
T3	PCM/TDM	672	500	46.304 Mbps
T4M	PCM/TDM	4032	500	
T5	PCM/TDM	8064	500	560.16 Mbps
			* requires regenerative receivers every 6000 ft	
			** requires special cable with low capacitance.	

1.7   Personal Communications Services (PCS)

Before the era of *personal communications services* (PCS), the cellular telephone used analog techniques. PCS uses digital communication modulation to transmit digital data utilizing wireless techniques. In order to enable multiple users in the same frequency band, several different multiple access techniques were implemented. They include *code division multiple access* (CDMA), *time division multiple access* (TDMA), and *frequency division multiple access* (FDMA), or a combination of any of these multiple access schemes.

Since the FCC allowed for companies to develop their own multiple access solution and chose not to implement a standard for the PCS development, interoperability became an issue. Two main implementations have emerged: CDMA and GSM. CDMA was adopted by several companies as a standard for sending and receiving digital communications. GSM was patterned after the standard in many European countries. Most PCS hardware accommodates either the CDMA or the GSM approach. Many PCS telephones are capable of using more that one technique. Also, other companies competing for the interoperability standard use TDMA and a combination of CDMA/TDMA.

Some systems have both PCS and cellular modes to allow the end user a choice and provide additional coverage and versatility. Since the onset of PCS hardware, the cellular industry also developed digital communication techniques to enhance the performance of cellular and compete with PCS. Some systems incorporate PCS, digital cellular, and analog cellular all in one handset.

Global System for Mobile Communications (GSM)

The GSM standard was developed in Europe; its standards are GSM-900 and the DCS-1800. Another popular standard set is *digital European cordless telecommunications* (DECT), which provides digital wireless communications and the DECT wireless telephone.

The US followed the standard set in Europe utilizing GSM technology at a slightly different frequency. The standard for GSM in the US is called PS-1900, which uses the GSM technology at 1900 MHz band.

Modulation for GSM

The modulation used for GSM systems is *Gaussian minimum shift keying* (GMSK). GMSK offers spectral efficiency properties with reduced sidelobes and the ability to send data at a high rate for digital communications. The bandwidth specified to send voice or data over the air is 200 kHz. This allows for multiple users to use the band at the same time, with minimal interference. This is called *frequency division multiple access* (FDMA), which means the users occupy a portion of the band that is free from interference. GSM also incorporates *frequency hopping* (FH) techniques, which are controlled by the base to prevent multiple users from interfering with each other.

The specification for GSM allows for the ability to send control bits to set the power output which also helps to prevent interference from other users since they should only use the minimum power that can make a reliable and quality connection. The power control helps the near-far problem, i.e., the distance between the handset and the base. If the handset is allowed to transmit at full power, then it will jam most of the other handsets. If power control is used, then the handset's power is reduced to the necessary power needed. This prevents jamming and saturation of the base and allows for handsets farther away from the base unit to turn up the power and reach the base. The optimum system would have the power level close to the same level regardless of the distance from the base station. The specification for GSM also allows adjustment of the duty cycle.

The data rate for GSM is approximately 277 kbps, with a 13 kbps vocoder. There are two types of GSM: Reflex 25, which uses two frequency channels, and Reflex 50, which uses four frequencies for twice the data speeds. The rate of change from one frequency to another is 2.4 kHz. DSPs are used to reduce the distortion in the overall system and detect the information reliably.

1.8   Cellular Telephone

Cellular telephones were originally analog frequency modulation systems operating in the 800 MHz band under the classification called AMPS. With the advent of PCS digital telephones, the cellular had to transition to digital modulation techniques to keep market place and compete with the PCS technology. Whereas cellular had the advantage of an infrastructure in place and antenna sites allocated, the PCS industry had to start from scratch. The cellular companies developed a digital modulation system and combined it with the analog technology, the dual mode AMPS. TDMA IS-54/IS-136 along with CDMA techniques were also implemented as digital cellular operation.

1.9   Industrial, Scientific, Medical ISM Bands

The FCC has designated frequency bands to be used by the industrial, scientific, and medical organizations. These bands are

13.533 – 13.567 MHz with the center frequency of 13.560 MHz
26.957 – 27.283 MHz with the center frequency of 27.120 MHz
40.66 – 40.70 MHz with the center frequency of 40.68 MHz
902 – 928 MHz with the center frequency of 915 MHz
2.4 – 2.5 GHz with the center frequency of 2.45 GHz
5.725 – 5.875 GHz with the center frequency of 5.8 GHz
24 – 24.25 GHz with the center frequency of 24.125 GHz

With the development of the PCS telephones, the FCC opened up the ISM bands for digital PCS communications with standards that have to be met for using these bands for wireless telephony communications. In addition, other applications, such as broadband communications and home networking are using the ISM bands for RF communications. The bands that were set aside and have become popular for use with RF solutions include

902 – 928 MHz
2.4 – 2.5 GHz
5.725 – 5.875 GHz

Currently, higher frequency bands, called *fixed wireless systems,* are being considered for bringing the signals to the home. Some of these systems use the ISM band from 24 – 24.25 GHz.

1.10 Summary

With the onset of the Internet and the desire to receive digital data at very high data rates, the telephone and connections are rapidly changing. Several types of modems have been developed, each version with a little faster data rate. Recently, the increase in data rates has been exponential. New methods both wired and wireless have been developed to transfer information. T lines are capable of much higher data speeds than the standard PSTN line. Both cellular (analog and digital) and PCS wireless systems are used for wireless telephony. Increased uses for these technologies will be utilized to provide other types of information.

1.11 References

1. Fike, John L., George E. Friend, Stephen J. Bigelow. *Understanding Telephone Electronics*. Indianapolis: SAMS Publishing, 1996.

2. Bullock, Scott R. *Transceiver System Design for Digital Communications*. Atlanta: Noble Publishing, 1995.

3. Tomasi, Wayne. *Advanced Electronic Communications Systems*. 4th ed. Englewood Cliffs, N J: Prentice Hall, 1998.

4. Papir, Zdzislaw and Andrew Simmonds. "Competing for Throughput in the Local Loop." *IEEE Communications Magazine* (May 1999).

2

High-Speed Modems

Broadband communications and home networking require higher data rates, which in turn demand higher speed modems. Several technologies are being developed to handle these increased data speeds. Only recently, V.34 modems were considered high-speed modems, delivering data rates up to 33.6 kbps. A modem speed of 28.8 kbps was considered very good. Then, the 56 k modems were introduced, with an average modem speed in the high 40 kbps. The tremendous increase in Internet use and the demand for very high-speed modems resulted in ISDN delivering 64 kbps, or combined 128 kbps. The demand kept forcing the speeds even higher as the DSL modems reached speeds in the Mbps.

When broadband communications and home networking became popular, speeds into the 10 and 11 Mbps range were possible, using parallel techniques, such as OFDM and more complex digital modulation schemes, for example, QAM systems. Data rates are continually improving: speeds up to 50 MHz, 100 MHz, 155 MHz, and even data rates into the Gbps are being developed.

2.1   Integrated Services Digital Network

Integrated services digital network (ISDN) has provided high-speed data rates for several years. ISDN uses the switched telephone network, the same as is used for the standard telephone. ISDN provides the user with two 64 kbps channels, also called *bearer channels* or *B-channels* and a lower speed D-channel. ISDN offers both high quality voice and high-speed data rates by combining the two 64 kbps channels to a 128 kbps bi-directional data channel. Although ISDN did not become the most popular option, it remains a viable solution for delivering high-speed information to the home. However, with the onset of other technologies providing much higher data rates, the ISDN market may remain small.

2.2   Digital Subscriber Loop

The *digital subscriber loop* (DSL) modems increase the digital data speeds over the standard telephone lines. The data rates for DSL type modems range from around 150 kbps up to nearly 7 Mbps. These types of modems are either symmetrical or asymmetrical. The symmetry refers to the data speed to the subscriber (downstream data) and the data rates back to the CO (upstream data). The downstream data is usually more important, since the end user is more concerned about the amount of time to download a file or information, especially for large files, than about the amount of time it takes to send data. The data rates, however, are inversely proportional to the distance from the CO. DSL modems use *frequency division multiplexing* (FDM) for up and down streams of data. To enhance the design for multiple users further, echo cancellation techniques allow the bands to overlap with minimal distortion.

Modulation

Four types of modulations are commonly used for DSL modems. They include, but are not limited to, the following:

1. *Carrierless amplitude/phase* (CAP) modulation was introduced to support HDSL. CAP is very similar to QAM. QAM uses both phase and amplitude variations to encode the data. QAM systems are capable of sending more data than standard QPSK systems since they not only encode the data using four different phase states, but they also include different amplitude levels in their encoding scheme (see Chapter 3). Therefore, instead of two bits of digital data per symbol sent, QAM provides three bits of digital data for every symbol transmitted. The main difference between QAM and CAP is that QAM is used to send the carrier and signal information whereas CAP only sends the data directly, without using a carrier.

2. *Discrete multitone* (DMT) modulation is the standard for *asymmetric DSL* (ADSL) encoding. DMT sends parallel channels of data at once to increase the overall throughput or payload of the data rate. If echo cancellation is used, then these channels can be closer together without

incurring problems of adjacent channel disturbances and more channels can be used in a given bandwidth to increase the overall data rate. The actual modulation in this system generally uses QAM along with this parallel technique to provide a high data rate system.

3. *16-QAM* is used to increase the data rates for a given bandwidth. With this type of modulation, there are more bits encoded for every symbol sent because of the increased phase and amplitude possibilities. Other QAM systems are utilized in communications systems and, as the technology increases, 32 and 64-QAM are possibilities for future systems.

4. *Multicarrier modulation* (MCM) modulates multiple carriers for sending information.

Error Correction Techniques

There are many types of error correction techniques that can be used to correct the errors that digital signals encounter in a typical transmission system. One such technique *Reed Solomon Forward Error Correction* (RS-FEC). The RS-FEC is known as a block code and shown in Figure 2-1a.

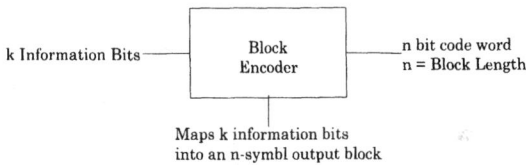

a. Block Codes (RS-FEC)

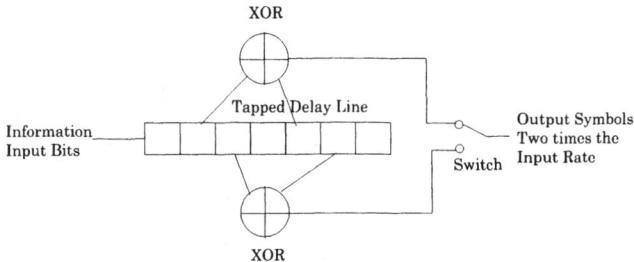

b. Rate ½ Convolutional Encoder

Figure 2-1 a. Block code, with RS-FEC. b. Convolutional code generator.

Block codes use lookup tables to determine the data that is sent. The data information is put into to the code block. Depending on this data, an output is determined by using the lookup tables (see Figure 2-1a). This type of error correction does not depend on previous data sent, i.e., it does not have any "memory" in its process. This signal is sent to the receiver where it decodes the data.

Another commonly used code for error correction the *convolutional code*. The convolutional encoder accepts the information input bits and sends them through a tapped digital delay line. As these input bits travel down the delay line, certain taps in the delay line are selected and the outputs of these are exclusive-or'd together to make two different data streams, as shown in Figure 2-1b. These data streams are multiplexed together, alternating between the data streams to produce one serial digital data stream at a rate of two times the input rate. This example has a rate of one half, since the input data is half the rate of the output data. There are also many other rates that can be utilized to generate a convolutional encoder as well as different length tapped delay lines. This example uses a 7-tap delay line called a constraint length of seven. Therefore, the proper reference to this example is a rate half convolutional encoder, constraint length 7.

Interleaving is used to minimize the effects of burst or pulse type jamming and noise signals. Interleaving delays and multiplexes data bits so that if a burst of signal or noise corrupts a section of data bits, they are not all from the same code word, which allows error correction to correct the bits in error (see Figure 2-2).

For the example, three messages are sent, with the criteria that the forward error correction can correct 1 bit error per message. Three messages are sent: 110, 001, 111. If a burst jammer is present and causes the three bits in the last message in error, then the last message (111) would be lost since the FEC is only capable of correcting 1 bit/message (see Figure 2-2). However, if interleaving is used with these three messages, the interleaved bit stream that is sent is 101 101 011. The burst jammer would corrupt the last three bits 011. Since interleaving is used, these three bits are the last bit of each of the messages. Therefore, since the FEC can correct 1 bit/message, then all of the messages are recovered (see Figure 2-2). Although this is a very simplified example of interleaving, the concept still applies for more sophisticated interleaving and wider spreading of the interleaved bits.

Interleaving and de-interleaving the data is done extensively in systems that are susceptible to burst jammers or burst noise. Commonly, when a burst occurs, the data bits at the time of the burst are corrupted. Therefore, the burst will cause errors in trying to receive these bits reliably. However, as the example shows, instead of corrupting 3 bits of error in one word, which would prevent the error correction from correcting that many errors for a particular digital word, interleaving provides a means that only one error per word occurs during the burst and the FEC can correct that error (see Figure 2-2).

Message 1 Message 2 Message 3

110   001   111

110 001 XXX

Burst Jammer

Message 3 is unrecoverable

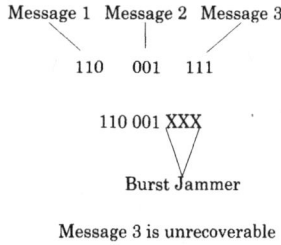

Interleaving

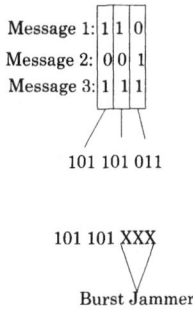

Message 1: 1 1 0
Message 2: 0 0 1
Message 3: 1 1 1

101 101 011

101 101 XXX

Burst Jammer

De-Interleaving

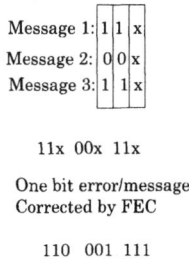

Message 1: 1 1 x
Message 2: 0 0 x
Message 3: 1 1 x

11x 00x 11x

One bit error/message
Corrected by FEC

110   001   111

Figure 2-2 *Interleaving is used against burst jammers/noise.*

Another error correction technique uses *Trellis coded modulation* (TCM). TCM combines QAM and error correction bits for sending the data information or payload. This error correction called *convolutional coding* (see Figure 2-1 b). Unlike block codes, it uses previous information to determine the next code state. Therefore, it is known as a "memory" system because it needs to store or have past knowledge of the previous code value. The receiver uses a Viterbi decoder for error correction and predicts the most likely path through the Trellis diagram, as shown in Figure 2-3.

000 001 010 011 100 101 110 111

000 001 010 011 100 101 110 111

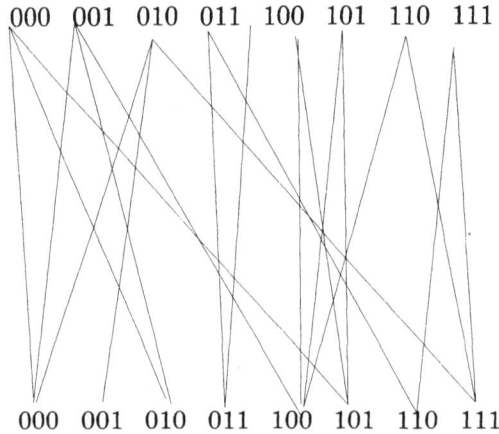

Figure 2-3 *Trellis coded modulation TCM diagram.*

These techniques are used to provide optimal performance, data reliability and low probability of error. Error correction schemes require overhead, which reduces the overall data rate throughput.

Digital Subscriber Loop Access Multiplexer (DSLAM) and PSTN Splitter

In order for the CO to handle multiple DSL signals and route them to the *Internet service provider* (ISP) or data network, a *DSL access multiplexer* (DSLAM) is required. The DSLAM accepts all of the DSL connections and combines or multiplexes them into one or a few links and sends this to the ISP or data network (see Figure 2-4).

A DSLAM typical implementation can handle 8,000 connections and provides data rates up to 1 Gbps. If all of the lines are active, then the maximum data rate for each user is 125 kbps.

Since DSL signals coexist with the telephone signals on the same PSTN or POTS line, a splitter is required to split the telephone signals from the DSL data signals (see Figure 2-4). This allows the end user to receive the telephone signals separately from the DSL data signals, which means that the subscriber can use both services at the same time.

Another type of DSL that adjusts or adapts the data rates according to the signal quality of the receiving signals is known as a *rate adaptive DSL* (RADSL). This allows for reliable reception because the signal is degraded so that several bit errors are received, which means the system can automatically adapt to a slower data rate. RADSL provides reliable detection with the optimal data rate for any given bit error rate threshold. Therefore, the more noise in the system, the more bit errors, which means the system lowers the data rate to provide an acceptable bit error rate. If the system noise is reduced, then the data rate increases to the acceptable bit error rate for the system.

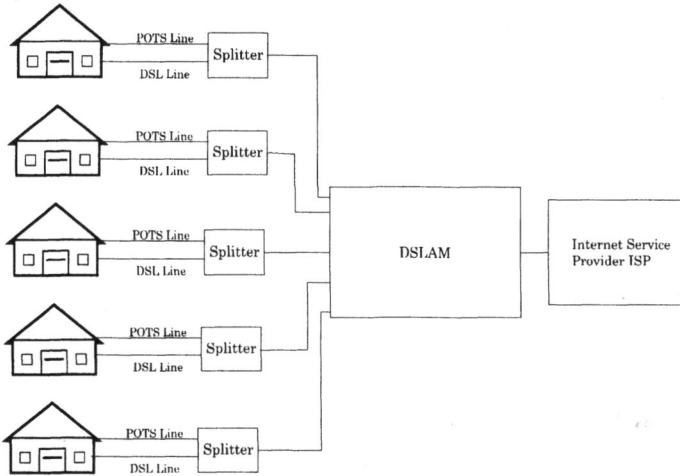

Figure 2-4 DSLAM and splitter connections for DSL Service.

2.3   Asymmetric Digital Subscriber Loop (ADSL)

Asymmetric digital subscriber loop (ADSL) is called asymmetric because its information is provided at a higher rate in one direction than in the opposite direction. Internet access needs high-speed downstream information for downloading large files off the Internet or collecting information on a particular Web site. There are different speeds for ADSLs available. One of the types of ADSLs provides a downstream data rate of 6.8 Mbps and an upstream data rate of 640 kbps. The range of this system is reliable up to 12,000 feet from CO or carrier service area CSA. Another type of ADSL provides a downstream of 1.544 Mbps and an upstream of 176 kbps, which provides an extended CSA. The lower speeds allow more uses to access the ADSL modem link; therefore, providers will often implement the lower speed capabilities to provide more end users with high-speed data. Some providers feel that it is better to have 1.5 Mbps capability to 85 percent of the end users, than to have the capability of 6 Mbps to 55 percent of the end users.

Discrete Multitone

In 1995, *discrete multitone* (DMT) was established as the official standard line code for ADSL. DMT divides the 1.1 MHz of spectrum into 256 discrete subchannels that are 4.3125 kHz each (see Figure 2-5).

Guard Band	Upstream	Downstream	Guard Band
6 Channels	32 Channels	Echo Cancellation/No Echo Cancellation	Echo Cancellation/No Echo Cancellation
		212 Channels/180 Channels	6 Channels/38 Channels
1-6	7-38	39-250/39-218	251-256/219-256

Each Channel = 4.3125 kHz   Total Bandwidth = 1.1 Mhz

Figure 2-5 *Discrete multitone DMT spectrum.*

The first six channels on the bottom end of the spectrum are not in use and only provide a guard band so that ADSL does not interfere with the PSTN signal. The next 32 channels allocated after the first six channels from channel 7 to channel 38 are for the upstream data information. Since the upstream data rates are not required to be as fast as the downstream data stream, fewer channels are allocated for the upstream. The number of downstream channels is dependent on whether or not echo cancellation is used. There are 212 channels allocated for the downstream information using echo cancellation. They range from channel 39 to channel 250 leaving six guard band channels on the upper end to prevent interference to other devices and ensure that the upper band does not have signal content in the band that was not allocated for ADSL. If echo cancellation is not used, then there are only 180 channels allocated for the downstream, which leaves 38 unused channels on the upper end of the band (see Figure 2-5).

Each channel transmits digital information using QAM modulation. Multichannel is used so that bad channels can be dropped. Also, DMT has the ability to change data rates for each of the channels separately. In case one channel is noisy, the data rates can be slowed down to achieve reliability.

ADSL was redirected to transport *Internet packets* (IP) to support variable length frames (Ethernet MAC), then ATM became popular with the G.Lite technology. With ATM over ADSL, users are connected to the network service provider via virtual circuits, called *private virtual circuits* (PVC). ATM contains dynamic rate adaptation so it can change the data rate dependent on conditions of the communication channel. Methods to map user data to the ADSL physical layer are

1. Integrated *network interface card* (NIC) used for G.lite.

2. Single user (via bridging), including Ethernet 10BaseT, *universal serial bus* (USB) and ATM25.

3. Multiple users (via routing), including twisted pair, wireless home networks, Ethernet 10BaseT and IEEE1394.

ADSL Lite — G.Lite

ADSL lite or G.Lite is a splitterless ADSL initiative from ITU G.992.2. The G.Lite modem technique eliminates the need for a splitter and the subscriber site, but all telephones and telephone products are required to have a filter to prevent the data from interfering with the telephone products. The data rates for G.Lite are 1.5 Mbps maximum for the downstream and 512 kbps maximum for the upstream. G.Lite systems are popular because the splitter is not required, which makes this system somewhat less expensive than others.

ADSL/ISDN

Another technology using both ADSL and ISDN is called ADSL over ISDN (AOI). The ADSL/ISDN system provides higher speeds and preserves the features of ISDN. The frequencies of operation for ISDN and ADSL are different: the frequency of operation for ISDN is up to 80 kHz or 120 kHz, depending on which ISDN type is used. ADSL, on the other hand, uses frequencies from 26 kHz to 1.1 MHz. ADSL occupies 256 channels that are approximately 4 kHz wide. This allows for a guard band between the PSTN rolloff at 4 kHz and the starting of ADSL at 26 kHz and allows enough bandwidth for the splitter filter to provide enough rolloff to prevent interference between the PSTN voice band and the ADLS data signals (see Figure 2-6).

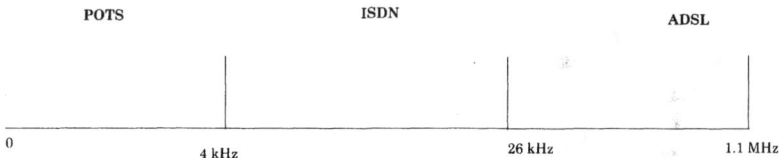

Figure 2-6 Spectrum of ADSL, ISDN, and PSTN.

Since ISDN and ADSL operate in two different bands, one of the technologies needs to move into the other's bandwidth or a completely separate bandwidth must be used. To obtain another completely different band allocation would be impractical.

To move one technique into the other's bandwidths seems much more practical. One solution could be to move the ISDN band over to the ADSL band. This move might cause problems in latency of processing the ISDN signal, which means it would not comply with the ISDN standard. Another possible problem would be that the power requirements of an ADSL modem might result in the line power not being capable of powering the device, which means that a power failure would cause the lifeline to be inoperable.

Another solution could be to move the ADSL to the ISDN band. The main problem with this solution would be that the ADSL operating in the new band would not be interoperable with the existing ADSL systems. However, this second approach is probably the best solution considering the serious problems associated with the first solution.

2.4   High-Speed Digital Subscriber Loop (HDSL)

High-speed DSL consists of three different types of standards: HDSL, HDSL2 and SDSL (see Table 2-1).

HDSL is used to provide T1 transmissions over existing twisted pair without requiring the setup. HDSL covers a range of up to 12,000 ft., which is over 2 miles. This HDSL is symmetric and either runs 1.536 Mbps, which requires 2 pair of wires, or 2.048 Mbps, which requires 2 and 3 pair of wires that are full duplex systems.

HDSL2 runs at a data rate of 1.536 Mbps, with only one pair of wires and provides full duplex, frequency division duplexing FDD, echo canceling, and trellis coding. HDSL2 covers a range of up to 18000 feet, which is greater than three miles.

SDSL stands for symmetric DSL, which is a variation of HDSL. The standard requires one pair of wires with data rates from 144 kbps to 1.5 Mbps. It also includes rate adaptation where the data rate will increase or decrease, depending on the line condition.

Another terminology that combines ISDN and DSL is IDSL. IDSL is symmetrical and capable of data transfers of 144 kbps in both directions.

2.5   Very High-Speed Digital Subscriber Loop (VDSL)

Very high-speed digital subscriber loop (VDSL) uses a combination copper wire and fiber optics, hybrid fiber/copper. Fiber optics is used from the CO to a fiber optic junction point (which is used for distribution to multiple homes). Copper wire runs from the fiber optic junction point to subscriber or home.

DMT modulation, or a variant of DMT called *zipper DMT*, can both eliminate certain frequency bands that cause RF interference. If QAM or CAP modulation is used, notch filters can reduce RF interference.

VDSL is capable of very high data rates due to the fiber optic cable. Two modes of data speeds are from 6.5 to 13 Mbps, and from 26 to 52 Mbps. Since data rates are relative, these data speeds may be considered low data rates as the speeds increase to 150 Mbps, 500 Mbps and even up to 1 Gbps (see Table 2-1). VDSL has been developed using both symmetric or asymmetric modes; however, for home, Internet, home shopping, and video on demand applications, asymmetric mode is generally used because the download direction is more demanding for the high speeds. VDSL can also share the same telephone line with telephone products similar to ADSL.

Table 2-1 *Summary of DSL modems.*

Modem Type	Downstream Speed	Upstream Speed
ADSL	1.544 Mbps	640 kbps
	6.8 Mbps	640 kbps
ADSL lite/G.lite	1.5 Mbps	512 kbps
ADSL/ISDN	1.544 Mbps	640 kbps
	6.8 Mbps	640 kbps
IDSL	144 kbps	144 kbps
HDSL	1.536 Mbps	1.536 Mbps
	2.048 Mbps	2.048 Mbps
HDSL 2	1.5236 Mbps	1.5236 Mbps
SDSL	144 kbps – 1.5 Mbps	144 kbps – 1.5 Mbps
VDSL	6.5 Mbps – 13 Mbps	6.5 Mbps – 13 Mbps
	26 Mbps – 52 Mbps	26 Mbps – 52 Mbps
Future Modems	150 Mbps, 500Mbps, 1Gbps	150 Mbps, 500Mbps, 1Gbps

2.6 Cable Modems

Cable modems are used with the coaxial cable to bring not only cable TV but also high-speed data. The existing infrastructure is already in place, so cable modems were developed to use the same coax TV cable to send and receive data for multiple purposes, including the Internet connection. However, the existing cable television systems must be upgraded for the more sophisticated and high-speed connections. Since high-speed data rates are desired, new cable is being used for both the data speeds and reliability. The new cable uses a hybrid of fiber optics and coaxial cable referred to as *hybrid fiber coax* (HFC), which is similar to the VDSL hybrid fiber copper. The difference is that cable modems use coax for the final connection and VDSL uses copper. Fiber optics is used from the CO to a fiber optic junction point, which is used for distribution to multiple homes. Coax cable is used from the fiber optic junction point to subscriber or home.

Cable modems link to the computer via Ethernet. As with most network systems, speed will be reduced, depending on the number of users. The bandwidth allocated for cable modems is 330 MHz or 450 MHz, with HFC delivering 700 MHz, which is shared. This provides Internet speeds to the end user of up to 10 MHz downstream and up to 2 MHz upstream. The cable modem connects the computer to the cable TV service (see Figure 2-7).

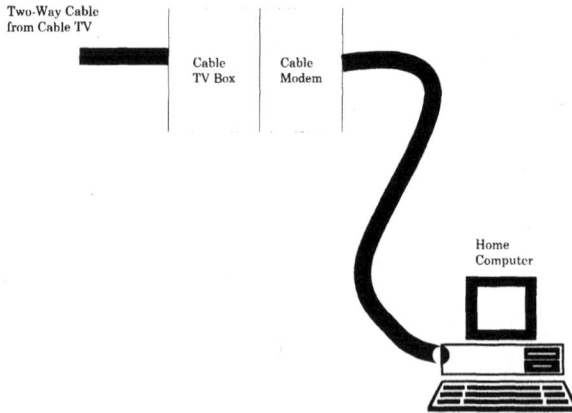

Figure 2-7 *Cable modems connected to the cable TV line.*

The cable head end distributes cable to the cable modem boxes at the end users site. Most of these cable modems are connected to a PC via an Ethernet 10BaseT or USB.

Several different types of modulation techniques are used in cable modems. They include QPSK, 64 QAM, 128 QAM, and even 256 QAM. Downstream data speeds reach to approximately 30 Mbps using 64 QAM which can be increased to nearly 40 Mbps using 256 QAM. The upstream data speeds range from 500 kbps up to 10 Mbps using QPSK and 16 QAM, depending on the bandwidth allocated.

One of the advantages of cable modems is that they are always on-line. There is no dial-up and receiving busy signals as it is the case with the PSTN modems. Also, the cable modem does not tie up the telephone line.

Cable modems also share the network amongst other users. The bandwidth is on demand, which means that even though multiple users are all on-line, the ones with the demand for high-speed data use the bandwidth when needed, which allows multiple users to utilize high-speed data, in bursts of time. If all the users at the same time demand the high-speed data, then the network slows down. This problem is solved by probabilities; the probability that all the users demand high-speed data at the same time is small. Adding more users to a system may require an additional amount of bandwidth or a redesign of the wiring system so that the network does not get overloaded.

2.7   Video Standards

Video distribution and networking is becoming important with the new applications and signals coming into the home. The video signals are included in the overall signaling and distribution in the home. Even the Internet connections are using video communications. Several video standards have been established. The following are the current standards, which may be modified as technology changes:

1. MPEG1 is mainly used for CD-ROM applications and delivers up to 1.5 Mbps.

2. MPEG 2 deals with compressed digital video and delivers the quality need for television and HDTV. It also is used to specify DVD systems.

3. MPEG 4 uses a special video coding that is object-oriented based. This makes it much easier to edit the objects in a picture than in a frame-oriented structure of the MPEG 1 and the MPEG 2 systems. Instead of editing the whole frame, the MPEG 4 system allows for editing just an object located in the frame. This becomes very useful for editing video.

2.8   Summary

High-speed modems and standards have been developed to meet the user demands for high-speed data information and faster communications with the Internet. DSL modems have been successful in reaching a large population because of the low cost structure, reliability, and high-speed data rates. ADSL and versions of ADSL, such as G.lite addressed a need in the market for many users. Hybrids using ISDN and ADSL have been produced, and high speed DSL such as HDSL and VDSL are reaching even greater modem speeds. Cable modems are rapidly becoming a solution for high-speed information to the home and office. Many new standards for bringing voice, data, and video to the home are being established as higher speeds and the number of users continue to grow.

2.9   References

1   Cooper, Ian R. and Mick A. Bramhall. Fujitsu Europe Telecom R&D Centre Ltd. "ATM Passive Optical Networks and Integrated VDSL," *IEEE Communications Magazine* (March 2000).

2   Issa, Jacques and Roy Bieda. "The G.DMT and G.Lite Recommendations, Part 2." *Communications Systems Design* (June 1999).

3   Chen, Walter Y. "The Development and Standardization of Asymmetrical Digital Subscribe." *IEEE Communications Magazine* (May 1999).

4   Kwok, Timothy C. "Residential Broadband Architecture Over ADSL nd G. Lite (G.992.2)." *IEEE Communications Magazine* (May 1999).

5   Papir, Zdzislaw and Andrew Simmonds, "Competing for Throughput in the Local Loop." *IEEE Communications Magazine* (May 1999).

6   "Cable Modem FAQ," "Overview of Cable Modem Technology and Services," "Cable Data Network Architecture," *Cable DataCom News*, Kinetic Strategies Inc., Arizona (January 2001).

3

Digital Modulation Techniques

To send digital signals from a transmitter to a receiver, a carrier frequency is used to carry the digital information. This is accomplished by using the digital signal to modulate the carrier. The carrier is modulated by using amplitude, phase, frequency modulation, or frequency hopping, or a combination of these. Phase modulation is the most common way of digitally modulating the carrier frequency. In addition, spread spectrum is used to achieve process gain, or improved reception in the presence of other signals, such as jammers or noise. Spread spectrum systems use more spectrum than is needed to send the digital information directly. Therefore, they require more bandwidth to achieve jammer protection. The same type of digital modulation techniques are used in a spread spectrum system, but a pseudorandom code at a much higher bit rate is combined with the data before modulation.

Parallel techniques are used to increase the data rate of a system. The data is sent out at the slower rate on multiple channels, for example frequency channels, and the receiver on the other end combines all of the parallel channels for an increase in overall data rate. One common scheme that uses parallel techniques to increase data rates is *orthogonal frequency division multiplexing* (OFDM). This technique uses orthogonality to allow overlap of the frequency cells to maximize data rates within a given bandwidth.

3.1   Phase-Shift Keying (PSK)

Phase-shift keying (PSK) is a type of modulation where the phase of the carrier is shifted to different phase states by discrete steps using a digital sequence. This digital sequence can be either the digitized data or a combination of digitized data and a high-speed spread spectrum sequence. There are many different levels and types of PSK. Several techniques are presented in this section; for others that are not discussed in this chapter, the same basic PSK principles apply.

Binary Phase Shift Keying (BPSK)

The basic PSK is the *binary PSK* (BPSK). BPSK shifts the carrier 0 or 180 degrees in phase, depending on the input digital waveform. For example, a +1 gives 0 degrees phase of the carrier, and –1 shifts the carrier by 180 degrees. The input digital signal and the carrier frequency are fed into a phase shifter or a mixer. The output of the mixer is passed through a filter to eliminate the harmonics and other spurious mixer products. The output of the filter is the BPSK signal (see Figure 3-1). The filter is also selected so that it causes minimum *intersymbol interference* (ISI) of the digital waveform.

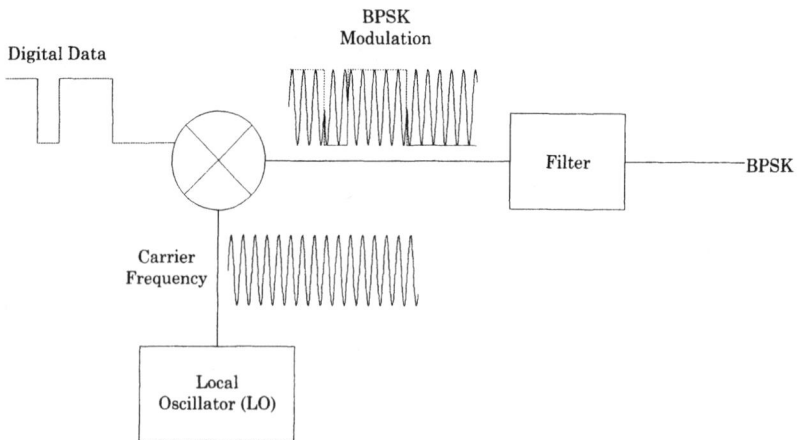

Figure 3-1 *Binary phase shift keying modulator.*

One of the simplest ways to view a PSK waveform is to use a phasor diagram. The phasor diagram shows the phase transitions of a PSK waveform. A BPSK phasor diagram consists of two different phase states, 0 and 180 degrees (see Figure 3-2).

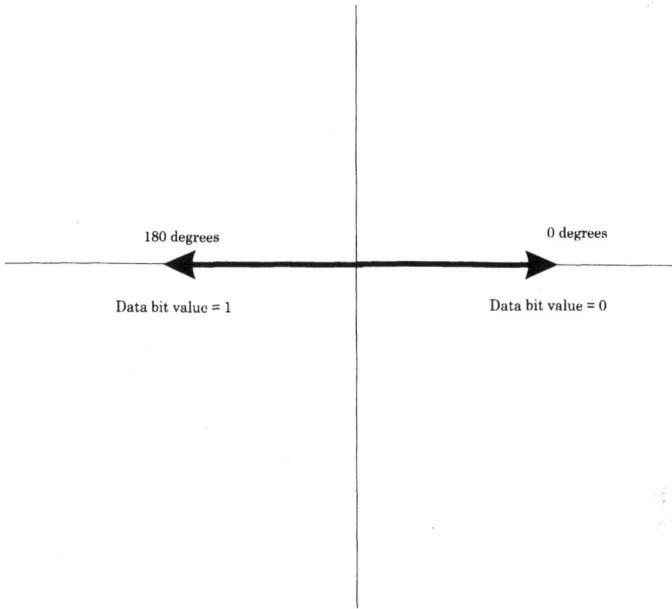

180 degrees   0 degrees

Data bit value = 1   Data bit value = 0

Figure 3-2 *Phasor diagram for a BPSK waveform.*

To produce the digital waveform, the data or information signal is digitized, encoded, and sent out in a serial bit stream. This serial bit stream then modulates the carrier, depending on the digital value. Therefore, the carrier is phase shifted between 0 and 180 degrees, depending on the input waveform. The LO is multiplied by a +1 or a –1 from the digital sequence, producing a 0 or a 180 phase shift.

The receiver detects the phase of the incoming signal and demodulates the carrier to produce the digital waveform. The demodulation takes on many different forms, which are determined by the complexity of the modulation scheme used. Some of the more common schemes used for BPSK are a squaring loop, a costas loop, or a delay loop, which is used for differential BPSK.

Differential Phase Shift Keying (DPSK)

The BPSK waveform in Figure 3-2 can be sent out as an absolute phase state, i.e. a 0 degree phase shift is a "1" and a 180 degree phase shift is a "0." This system is required to be coherent and maintain absolute phase through the system and in the receiver. Another way to perform this function is to use differential BPSK (DBPSK), which monitors the change of phase. That is, a change of phase (0 to 180 or 180 to 0) represents a "1" and no change (0 to 0 or 180 to 180) represents a "0." This scheme is easier to detect because the absolute phase does not need to be determined; only the change of phase is monitored. The phasor diagram for differential BPSK is the same as coherent BPSK, since it only shows the phase state.

Differential mode can be applied to various phase shifting schemes and higher order phase shift schemes. Differential mode results in some degradation compared to coherent PSK; however, it is dependent on the S/N or Eb/No level and the operating position on the probability of error curve. The differential scheme experiences a loss compared to coherent PSK systems because when a bit error occurs, the next bit is also affected since it relies on the differential value of the previous bit. Therefore, when a bit error occurs, the differential method experiences two bit errors.

Differential techniques can be applied in higher order PSK systems and include *differential quadrature phase shift keying* (DQPSK) and *differential eight phase shift keying* (D8PSK).

Quadrature Phase Shift Keying (QPSK)

QPSK modulation uses four phase states, 0, 90, 180, 270 to represent the data that is sent. Therefore, QPSK can send out twice as much data for a given bandwidth or symbol rate. The phasor diagram shows the four phase states and two bits of information for every symbol or phase shift. For example, 0 phase state represents two bits of data that are both zero value. (see Figure 3-3).

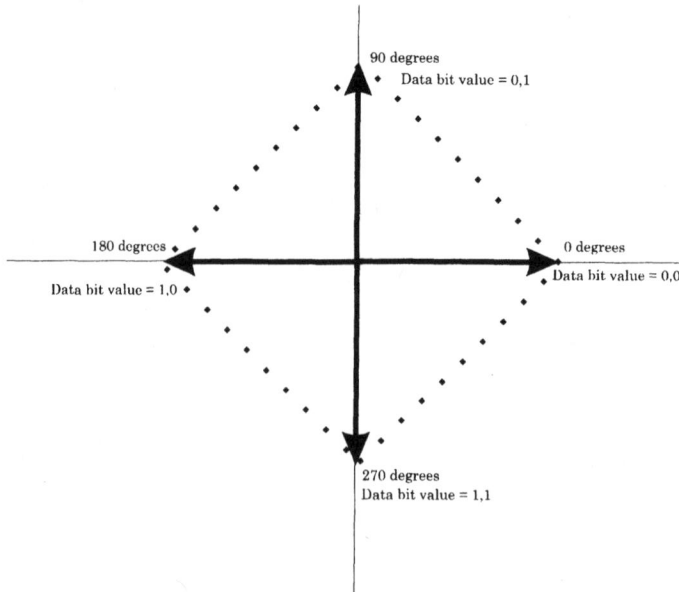

Figure 3-3 *Phasor diagram for a QPSK modulation scheme.*

Another application of QPSK is the ability to send out two separate BPSK messages, one using 0, 180 degree phase shifts called the I channel, and one using the 90, 270 degree phase shifts called the *Q channel*. Since they are orthogonal (see Chapter 4), the two different data streams can be recovered as separate messages.

A simple way of generating QPSK is shown in Figure 3-4. Two different digital streams are used for each of the BPSK sections. The carrier is split in quadrature by the hybrid so that the two BPSK signals are in quadrature, 0, 180 and 90, 270. The two BPSK signals are summed to produce the QPSK signal. The method of generation depends on the type of system used. The transitions of the two sequences occur at the same time to provide phase shifts of 0, 90, 180, and 270 degrees.

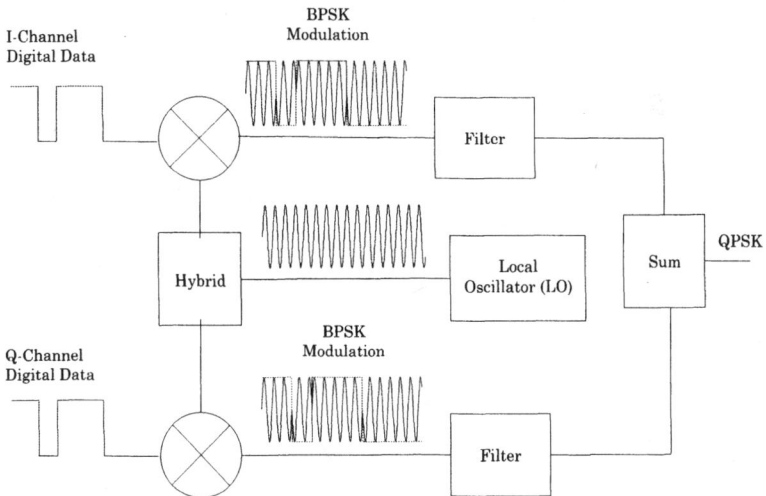

Figure 3-4 *QPSK generator.*

Differential QPSK can also be used. The detection or demodulation of the DQPSK is much simpler, with only a slight degradation in performance. The phasor diagram would be the same as in Figure 3-4. The generator would be changed slightly to look only at the change in the incoming bit streams for the I and Q channels and a carrier phase shift of 180 degree swould occur in either channel when it detects a change in bit value for that channel, for example between a "0" and a "1." If the bit value stays the same, regardless of the actual value, for example a "1" bit value to a "1," then there is no change in bit value and the phase of the carrier for that channel from one bit value to the next would not change. Therefore, there would be "0" phase shift for that channel.

Offset QPSK (OQPSK)

Another type of quadrature PSK is referred to as *offset QPSK* (OQPSK). This configuration is identical to QPSK except that one of the digital sequences is delayed by a half cycle so that the phase shift can only shift by 0 or 90 degrees and can only change to adjacent phasors (no 180 degree phase shift possible). A block diagram of an OQPSK generator is shown in Figure 3-5.

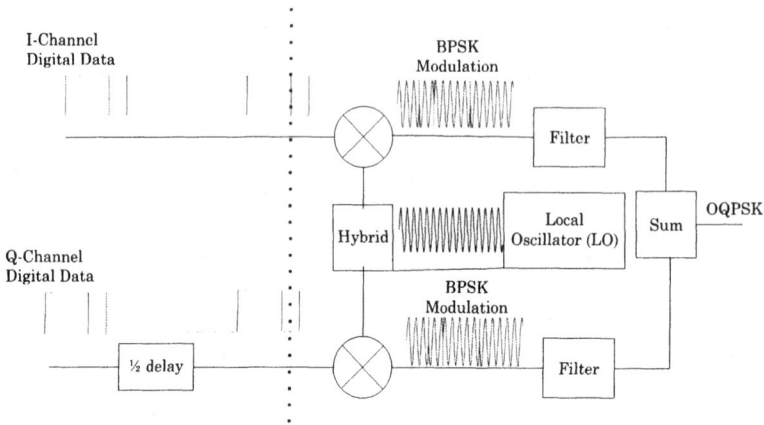

Figure 3-5 *OQPSK generator.*

The phasor diagram is identical to QPSK (180 degree phase shifts are not possible). This prevents zero crossover points and provides smoother transitions with less chance of error and reduces the amplitude modulation effects common to both BPSK and QPSK. The zero crossover points means that during the transition from 0 degrees to 180 degrees the amplitude of the phasor goes to 0. Therefore, since the hardware does not allow for an instantaneous change in phase, the changing amplitude produces amplitude modulation. The OQPSK only changes 90 degrees so that the phasor only goes through a –3dB amplitude degradation point, which reduces the amplitude modulation.

Differential techniques can also be used with OQPSK, which operates similar to DQPSK. The difference in phase change would be +90, –90, and 0 or no change. A 180 degree phase change would not be possible with DOQPSK (see Figure 3-5).

8-PSK and D8PSK

Another type of PSK that uses eight different phase states is 8-PSK modulation. 8-PSK modulation uses the following eight phase states: 0, 45, 90, 135, 180, 225, 270, and 315, which represents the data that is sent. Therefore, 8-PSK can send out three bits of information for every phase possibility or symbol. The phasor diagram shows the eight phase states and three bits of information for every symbol or phase shift. For example, 0 phase state represents three bits of data that are all equal to zero value and so on (see Figure 3-6).

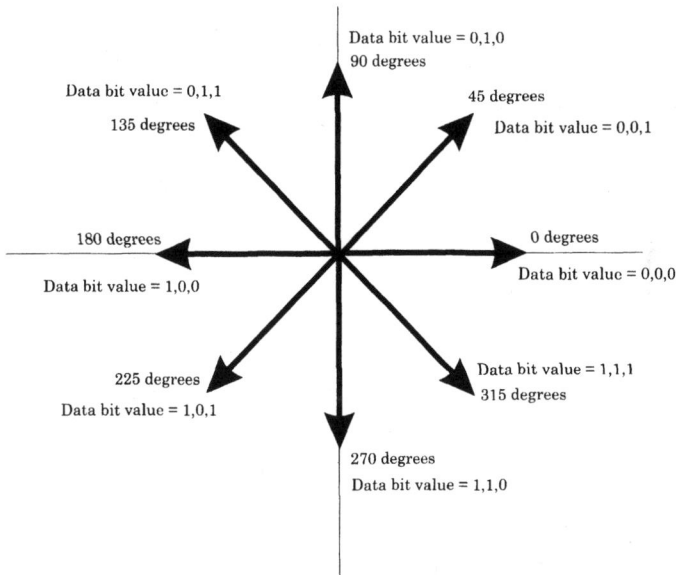

Figure 3-6 *8PSK/D8PSK phasor diagram.*

Differential 8-PSK (D8PSK) can also be used, which allows for a much simpler demodulation of the D8PSK and results in a slight degradation in performance. The phasor diagram would be the same as in Figure 3-6 but the actual phase change occurs with respect to the previous phase state.

Higher Order PSK

The above analysis can be extended for higher order phase-shift keying systems. The same principles apply; they are extended for additional phase states and phase shifts.

There are many other PSK configurations that are in use today. One of the main concerns using PSK modulators is the *amplitude modulation* (AM) that is inherent when phase states are changed. For example, BPSK switches from 0 to 180 degrees. Since this is not instantaneous due to bandwidth constraints, the phasor passes through zero amplitude in the process. The more bandlimiting, the more amplitude modulation there is in the resultant waveform. Various schemes have been developed to reduce this amplitude modulation problem. For example, OQPSK only allows a maximum phase shift of 90 degrees, which significantly reduces the AM problem.

Minimum Shift Keying

If the OQPSK signal is smoothed by sinusoidal weighting, another *minimum shift keying* (MSK) is produced. The weighting frequency is equal to half the chip rate, so that it smoothes the transition of a +1, −1 by one cycle of the sine wave weighting signal. This provides a smoothing of the 180 degree phase shifts that occur in both of the channels (see Figure 3-7).

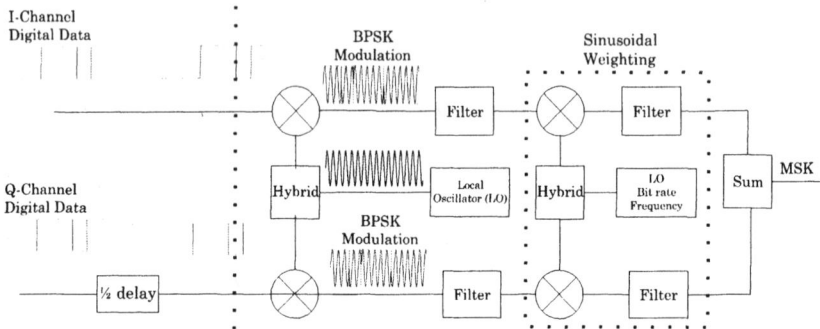

Figure 3-7 *MSK Generator using sinusoidal weighting of the I and Q channels.*

This smoothing slows the phase transitions of the two channels and reduces the high-frequency content of the spectrum that results in attenuation of the sidelobes that contain the high frequencie. Further, it minimizes the AM problem and makes the transition continuous, which prevents the −3 dB amplitude that is present in OQPSK. The phasor diagram shows that the transition follows a circle around the center with the same amplitude during the transition, hence no AM (see Figure 3-8). Therefore, the spectrum is said to be efficient compared to standard PSK systems (for example, BPSK and QPSK) since more power is contained in the main lobe and less in the sidelobes. MSK can be analyzed in two ways. The first is sinusoidal weighting of an OQPSK signal as discussed above.

Data bit value = 0,1
90 degrees

180 degrees
Data bit value = 1,0

0 degrees
Data bit value = 0,0

270 degrees
Data bit value = 1,1

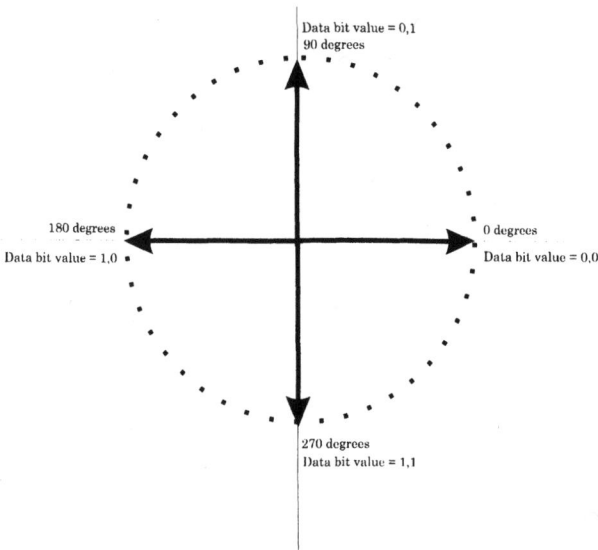

Figure 3-8 *Phasor diagram for MSK.*

Quadrature Amplitude Modulation (QAM)

Quadrature amplitude modulation (QAM) is very similar to QPSK, but includes amplitude variations for increase data capacity. QPSK has four phase possibilities so that each phase state represents two bits of information. If two amplitude states are added to the four phase states, the total number of possibilities is eight, i.e., for each state there are three bits of information (see Figure 3-9). More amplitude and phase states can be added to provide an increase in data rates for a given bandwidth.

QAM modulation schemes are used extensively in broadband communications and home networking to optimize the amount of data throughput along with other schemes to provide the optimal data rate for a given medium and bandwidth.

Differential QAM schemes are also used for simplicity in detection and the same principles apply to the QAM systems. For differential QAM, both the phase and the amplitude are differentially encoded so that for a change in either amplitude or phase, the bit information is detected.

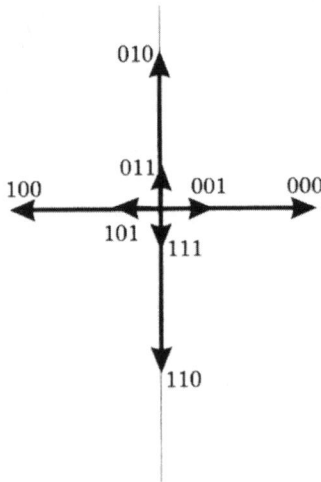

Figure 3-9 *QAM phasor diagrams.*

Side Lobe Reduction Methods

One of the problems with phase-shift keying and frequency shift keying systems is that the sidelobes can become fairly large and interfere with adjacent channel operation. The sidelobes continue theoretically to infinity. The main problem is usually associated with the first or second sidelobes, which are higher in magnitude. In order to confine the bandwidth for a particular waveform, a filter is required. The main problem with filtering a PSK or a FSK signal is that this causes the waveform to be dispersed or spread out in time. This can cause distortion in the main signal as well as is interference from adjacent pulses.

A Gaussian filter is one of the possibilities for reducing sidelobes. Another possibility is the *raised cosine filter,* which is designed by raising or adding a positive offset to a single cosine waveform that closely approximates the main lobe of the PSK waveform in the frequency domain (see Figure 3-10). A *raised cosine filter squared* is a squared raised cosine function. A Sin2X/X and a raised cosine, which is used as a filter for the PSK frequency domain signal are shown in Figure 3-10.

Raised cosine filter and raised cosine filter squared reduce the bandwidth of the signal but provide lower ISI than standard filters, such as Butterworth or Chebychev. Raised cosine filters are also used for better performance.

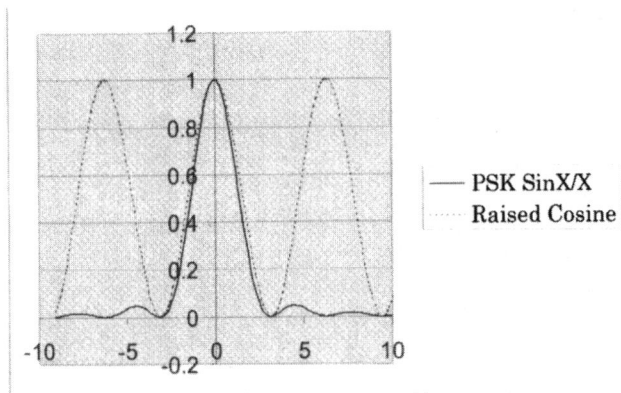

Figure 3-10 *Raised cosine filter used to reduce sidelobes of the PSK or FSK frequency domain signals.*

3.2   Frequency Shift Keying (FSK)

Another modulation scheme to send data is shifting between two differ-ent frequencies so that a +1 is frequency 1 and –1 is frequency 2. An alter-native method of developing MSK is by *frequency shift keying* (FSK) and set-ting the frequency shift rate equal to the frequency separation between the two frequencies. A simple way of generating MSK is shown in Figure 3-11.

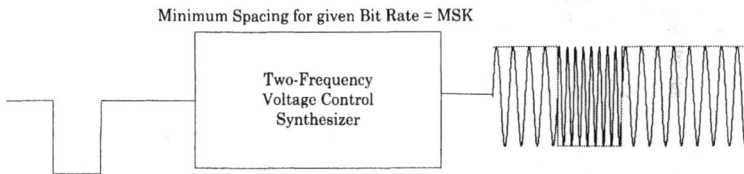

Figure 3-11 *MSK generator using FSK modulation.*

A two-frequency synthesizer is frequency-shifted by the binary bit stream according to the PN code. If the rate is set to the minimum chipping rate, then MSK is the resultant output. A simulation shows different types of FSK with different spacings with respect to the chip rate (see Figure 3-12).

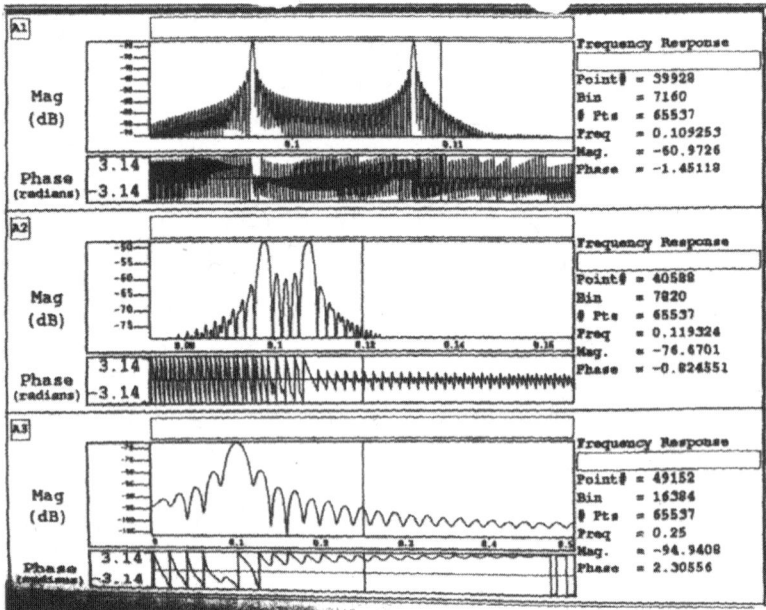

Figure 3-12 *FSK and MSK simulation in the frequency domain.*

If the frequency spacing is closer than the chip rate, the information cannot be recovered. If the spacing is too far apart, the information can be retrieved using FSK demodulation techniques but is not as band efficient as MSK since it takes more bandwidth for the same amount of data sent. As the frequency spacing approaches the chip rate, the resultant spectrum is MSK (see Figure 3-12).

Differential techniques can also be used for FSK. Instead of a data bit value "1" being one frequency and a data bit value "0" being the second frequency, a change in frequency is a "1" data value and no change in frequency is a "0" data value.

Gaussian Minimum Shift Keying (GMSK)

A Gaussian filtered version of MSK is a method called *Gaussian minimum shift keying* (GMSK). A Gaussian shape curve is the standard bell shape curve showing Gaussian distribution, which is used extensively in probability theory. The shape of this curve closely approximates the shape of the main lobe in the MSK waveform in the frequency domain (see Figure 3-13) because it has to pass the main lobe with minimal degradation and filter out the sidelobes. This provides effective use of the band and allows multiple users to coexist in the same band with minimum interference.

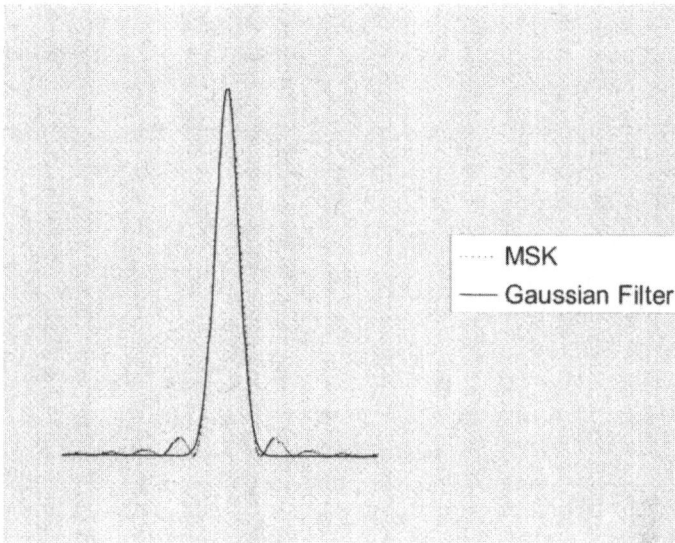

Figure 3-13 *A Gaussian filter is used to filter MSK to produce GMSK.*

Further, PSK systems can be combined with amplitude modulation to produce a hybrid called *quadrature amplitude modulation* (QAM). GMSK and QAM systems are very popular in the broadband communications and home networking arena where the data rates and the the efficient use of the bandwidth are very important. Standards in these areas are being set for standardization and interoperability.

3.3   Multiple Users

To prevent interference between systems for multiple users in the same geographical area, a multiple access scheme needs to be allocated Interference can be reduced using:

1.   Time Division Multiple Access (TDMA)

2.   Code Division Multiple Access (CDMA)

3.   Frequency Division Multiple Access (FDMA)

These techniques are shown in Figure 3-14.

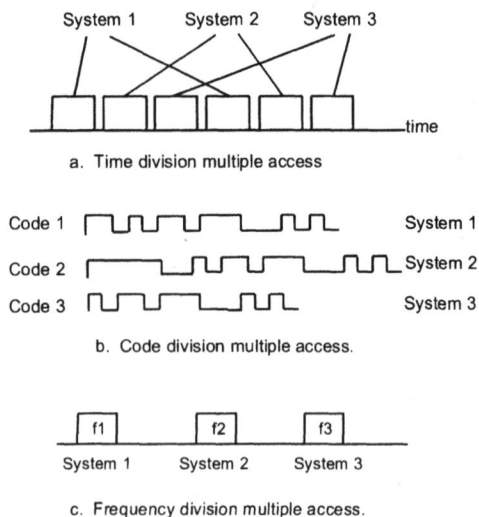

Figure 3-14 *Multiple access schemes for multiple users.*

Some control is required in all of these systems to ensure that each system in an operating area has different assignments.

TDMA provides interference reduction by having the systems communicate at different times or time slots at the same frequencies and codes (see Figure 3-14). If the systems have predetermined time slots, then they are considered to be *time division multiplexing* (TDM). If a system accesses the time slot, say on a first come first serve basis, then the system is a true TDMA system.

CDMA reduces interference because the systems communicate on different codes, preferably orthogonal codes or gold codes, at the same frequencies and time, which provides minimum cross-correlation between the codes resulting in minimum interference between the systems (see Figure 3-14). Generally, the shorter the codes, the more cross-correlation interference is present, and the fewer optimal codes can be obtained.

FDMA provides interference reduction by letting the systems communicate on different frequencies and possibly at the same time and code (see Figure 3-14). This provides very good user separation since filters with very steep roll-offs can be used. Each user has a different frequency of operation and can communicate continuously on that frequency.

Each of the multiple user scenarios discussed above reduces interference and increases the communications capability in the same geographical area.

3.4   Parallel Techniques to Increase Data Rates

Parallel techniques can be used to increase the data rate of a system. By sending data parallel, more data can be sent for a given amount of time. If the data rate is 1 Mbps, using ten channels can increase the data throughput to 10 Mbps.

One of the more common methods that use this type of parallel system to increase the data rate of a system is *orthogonal frequency division multiplexing* (OFDM).

3.5   Orthogonal Frequency Division Multiplexing (OFDM)

OFDM allows high-speed data to be transmitted in a parallel system. This is accomplished by taking the serial data stream from the transmitter and converting it to parallel streams. These streams are multiplexed and have different frequency channels in the OFDM system that sends the parallel streams of data to the receiver. The receiver then performs a parallel to serial conversion to recover the data. Also, the system can be made adaptive to only select the channels that are clear for transmission of data.

Another use for OFDM is an extension of FDM for multiple users. This technique uses the principle that if signals are orthogonal they do not interfere with each other (see Chapter 4).

For OFDM, there are multiple users for each of the frequency slots, and the adjacent slots are made to be orthogonal to each other so that they can overlap and not interfere with each other. This allows more users to operate in the same given bandwidth using overlapping signals (see Figure 3-15).

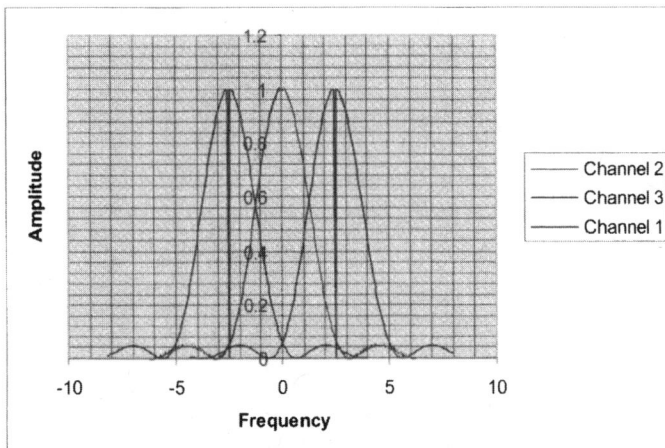

Figure 3-15 OFDM allows multiple users to operate in overlapping bands.

3.6   Spread Spectrum Systems

Many systems today us spread spectrum techniques to prevent other signals from jamming the reception. PSK technology can be used with a high-speed pseudo-random code to spread the bandwidth and provide process gain to receive the desired signal above the jamming signal. This is known as *direct sequence* (DS). The data is usually exclusive-or'd with a pseudo-random or *pseudo-noise* (PN) code that has a much higher chipping rate than data rate. This produces a much wider occupied spectrum in the frequency domain (spread spectrum), as shown in Figure 3-16.

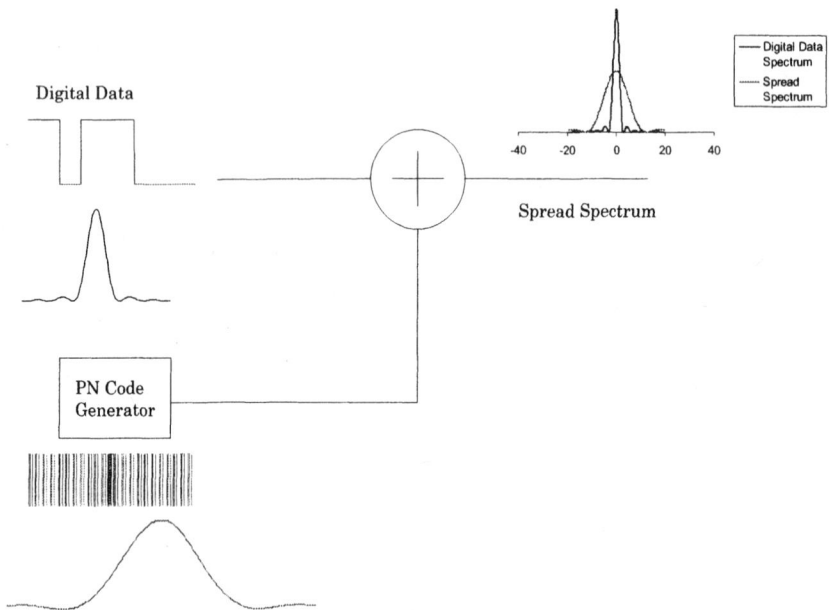

Figure 3-16 *Spread spectrum generator.*

Other forms of spread spectrum, including frequency hopping, time hopping, and chirped FM systems can be used to spread the signal by using more bandwidth than is required to send the data in order to improve the system against jamming type signals.

The receiver is designed to despread the spread spectrum signal, which at the same time, spreads the jamming signal. By selecting a filter for the desired despread (narrowband) signal, nearly all of the desired signal power passes through the filter, and only a small part of the spread (wideband) jamming signal makes it through the filter. Therefore, the signal-to-jammer ratio is improved by the spread spectrum process.

3.7   Summary

Digital modulation waveforms are used extensively in broadband communications and home networking. BPSK is the simplest form of phase-shift keying; it shifts the carrier 0 or 180 degrees. QPSK is used to increase data rates or to provide for multiple channels. Differential techniques are used to provide simpler demodulation methods and are acceptable with all types of modulations. OQPSK is used to minimize amplitude modulation by eliminating the 180-phase shift possibility and provides another method to generate MSK and GMSK. Higher order PSK systems can be analyzed much the same way only with more phase states and phase transitions. However, the more phase shift possibilities, the harder it is to detect and resolve the different phase states. Therefore, there is a limit on how many phase states can be sent for good detection. This limit seems to grow with better detection technology, but caution must be given to the practicality of how many phase states can be sent out for standard equipment. Amplitude modulation points are combined with phase states to produce a hybrid QAM to further increase the data capacity for the bandwidth channel. FSK is another method of encoding data using two frequencies. Minimum spacing provides another method of producing MSK.

TDMA, CDMA, FDMA are techniques to allow multiple users. Parallel systems can be implemented to increase the data rate, which is dependent on the number of parallel channels used. OFDM is a way to transmit parallel overlapping channels for bandwidth efficiency. OFDM can also be used to allow multiple users to operate in an FDM mode, with more users for a given bandwidth. Spread spectrum techniques are used to provide process gain to reduce the effects of jammers and allow more efficient use of the spectrum for multiple users.

3.8   References

1. Bullock, Scott R. *Transceiver and System Design for Digital Communications*. 2nd ed. Atlanta: Noble Publishing, 2000.

2. Inglis, Andrew F. *Electronic Communications Handbook*. New York: McGraw-Hill, 1988.

3. Crane, Robert K. "IEEE Transactions on Communications." vol. Com-28, no. 9 (September 1980).

4. Schwartz, Mischa. *Information, Transmission, Modulation and Noise*. New York: McGraw-Hill, 1980.

5. Haykin, Simon. *Communications Systems*. New York: John Wiley & Sons, 1983.

6. Holmes, Jack K. *Coherent Spread Spectrum Systems*. New York: John Wiley & Sons, 1982.

7. Bullock, Scott R. "Use Geometry to Analyze Multipath Signals." *Microwaves & RF* (July 1993).

8. Bao-yen Tsui, James. *Microwave Receivers with Electronic Warfare Applications*. New York: John Wiley & Sons, 1986.

9. Bullock, Scott R. "Phase-Shift Keying Serves Direct-Sequence Applications." *Microwaves and RF* (December 1993).

10. "Satellite Communications Handbook, Fixed Satellite Service," *International Telecommunications Union, International Radio Consultative Committee, Geneva* (1988).

4

Application and Uses for Orthogonal Signals

Orthogonal signals are used extensively in the communications industry. They range from a simple sine/cosine quadrature signals to multiple signals whose inner product is equal to zero. Orthogonal signals can be used for several different applications.

Quadrature signals can be used to send and receive separate information channels on each orthogonal signal with minimal interference between them. For example, if a QPSK modulator is used in a system, two different data streams, one for the I channel (0, 180 degrees) and one for the Q channel (90, 270 degrees), can be sent simultaneously and received on the other end as separate data streams.

Orthogonality can also be applied to polarizations in an antenna system. Two signals can be transmitted on different polarizations or two parallel channels can be used with the same frequency for increased data rates. The channels can use either horizontal/vertical polarization separation or the *left hand circular polarized/right hand circular polarized* (LHCP/RHCP) technique.

Orthogonal principles are also used to separate desired signals from jammers using a *Gram Schmidt orthogonalizer* (GSO). The GSO forces the desired signal to be orthogonal to the jamming signal so that it can be used to eliminate the jamming signal.

Another application for orthogonal signals is to prevent adjacent channel interference. This application allows for the system to be able to overlap multiple frequency channel signals with minimum interference between the channels and still guarantee reception and detection of phase-shift keyed signals. Thus, more data can be sent over a given bandwidth because the channels can overlap. The data is sent in parallel channels, and if they can be put closer together or overlapped, then more parallel channels can be used, which results in an overall higher data rate. Also, more users can operate in a given bandwidth due to the over lapping channels. In this scenario, each user occupies a channel, or multiple channels, and these channels can operate closer to each other without interference.

4.1   Quadrature Sine/Cosine Signals

Signals that are in quadrature, or on a simple term, 90 degrees out of phase from each other, are considered to be orthogonal. Suppose two signals are sent, $f_1(t)$ and $f_2(t)$ on a single carrier but 90 degrees out of phase, $f_1(t)\cos(\omega t)$ and $f_2(t)\sin(\omega t)$. Using a quadrature downconversion, $f_1(t)$ is retrieved in path 1 and $f_2(t)$ is recovered in path 2 (see Figure 4-1).

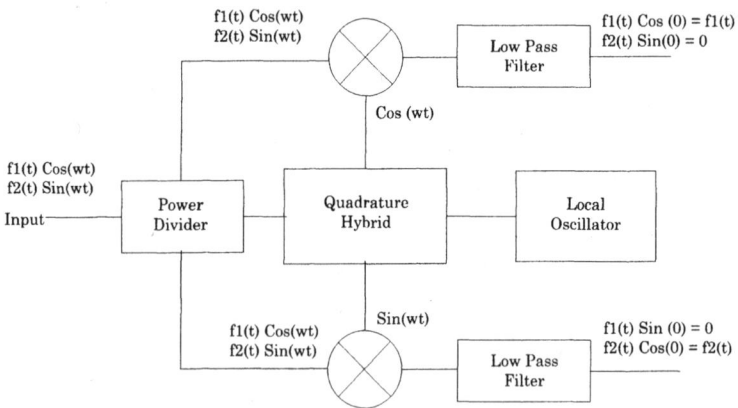

Figure 4-1 *Quadrature downconverter.*

Using the sum and differences that are produced by mixing or multiplying two signals together produces:

Path 1:

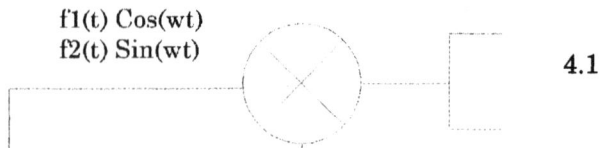

4.1

Path 2:

f_2 signal: $f_2(t)\sin(\omega t)*\cos(\omega t) = f_2(t)\left[\dfrac{1}{2}\right]\sin(\omega t - \omega t) + \dfrac{1}{2}\sin(\omega t + \omega t)$

4.2

$$= \dfrac{1}{2}f_2(t)\big[\sin(0) + \sin(2\omega t)\big]$$

f_1 signal: $f_1(t)\cos(\omega t)*\sin(\omega t) = f_1(t)\left[\dfrac{1}{2}\right]\sin(\omega t - \omega t) + \dfrac{1}{2}\sin(\omega t + \omega t)$

4.3

$$= \dfrac{1}{2}f_1(t)\big[\sin(0) + \sin(2\omega t)\big]$$

f_2 signal: $f_2(t)\sin(\omega t)*\sin(\omega t) = f_2(t)\left[\dfrac{1}{2}\right]\cos(\omega t - \omega t) + \dfrac{1}{2}\cos(\omega t + \omega t)$

4.4

$$= \dfrac{1}{2}f_2(t)\big[\cos(0) - \cos(2\omega t)\big]$$

By using a low-pass filter to eliminate ($2\omega t$) in the equations above, we obtain:

Path 1:

f_1 signal: $\dfrac{1}{2}f_1(t)\big[\cos(0)\big] = \dfrac{1}{2}f_1(t)[1] = \dfrac{1}{2}f_1(t)$ 4.5

$f_1(t)$ is retrieve at half the amplitude.

f_2 signal: $\dfrac{1}{2}f_2(t)\big[\sin(0)\big] = \dfrac{1}{2}f_2(t)[0] = [0]$ 4.6

$f_2(t)$ is not present.

Path 2:

4.7

$f_1(t)$ is not present

4.8

$f_2(t)$ is retrieve at half the amplitude.

Therefore, Path 1 only contains the f_1 signal and Path 2 only contains the

f_2 signal. Two signals can be sent on the same frequency but using orthogonal principles to retrieve both signals.

In digital modulation, a BPSK signal uses 0, 180 degrees to send information. A QPSK signal uses 0, 90, 180, 270 degree to send twice as much data on the same carrier. However, if two different data streams are needed to be sent on the same carrier, f_1 could be sent using 0, 180, and f_2 could be sent using 90, 270. Since these data streams are orthogonal, they can be retrieved using quadrature downconversion. Assuming the low pass filter is already in place, an example is as follows:

Path 1:

f_1 signal: $f_1(t)\cos(\omega t + 0,180) * \cos(\omega t) = f_1(t)\left[\dfrac{1}{2}\cos((\omega t - \omega t) + 0,180)\right]$

$$= \dfrac{1}{2}f_1(t)\left[\cos(0 + 0,180)\right]$$

$$= \dfrac{1}{2}f_1(t)\left[\cos(0,180)\right]$$

$$= +/-\dfrac{1}{2}f_1(t)$$

4.9

f_2 signal: $f_2(t)\cos(\omega t + 90,270) * \cos(\omega t) = f_2(t)\left[\dfrac{1}{2}\cos((\omega t - \omega t) + 90,270)\right]$

$$= \dfrac{1}{2}f_2(t)\left[\cos(0 + 90,270)\right]$$

$$= \dfrac{1}{2}f_2(t)\left[\cos(90,270)\right]$$

$$= 0$$

4.10

Therefore, the plus and minus digital data of $f_1(t)$ is retrieved in Path 1.

Path 2:

f_1 signal: $f_1(t)\cos(\omega t + 0,180) * \sin(\omega t) = f_1(t)\left[\dfrac{1}{2}\sin((\omega t - \omega t) + 0,180)\right]$

$$= \dfrac{1}{2}f_1(t)\left[\sin(0 + 0,180)\right]$$

$$= \dfrac{1}{2}f_1(t)\left[\sin(0,180)\right]$$

$$= 0$$

4.11

f_2 signal: $f_2(t)\cos(\omega t + 90,270) * \sin(\omega t) = f_2(t)\left[\dfrac{1}{2}\sin((\omega t - \omega t) + 90,270)\right]$

$$= \dfrac{1}{2}f_2(t)\left[\sin(0 + 90,270)\right]$$

$$= \dfrac{1}{2}f_2(t)\left[\sin(90,270)\right] \qquad\qquad 4.12$$

$$= +/-\dfrac{1}{2}f_2(t)$$

Therefore, the plus and minus digital data of $f_2(t)$ is retrieved in path 2. A phasor diagram showing the two orthogonal data streams are shown in Figure 4-2.

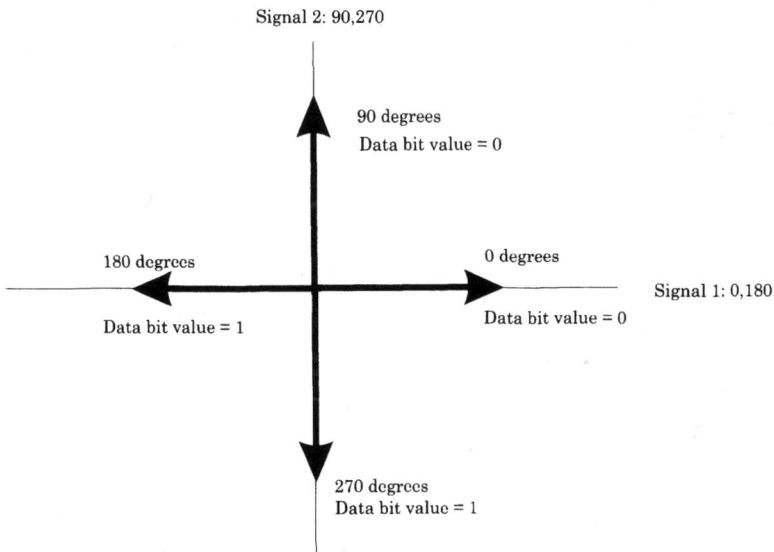

Signal 2: 90,270

90 degrees
Data bit value = 0

180 degrees

0 degrees

Signal 1: 0,180

Data bit value = 1

Data bit value = 0

270 degrees
Data bit value = 1

Figure 4-2 *Phasor diagram showing two orthogonal data streams.*

Quadrature downconversion also ensures that the data is recovered regardless of the phase ambiguity of the incoming signal. This is accomplished by a costas loop (see Figure 4-3).

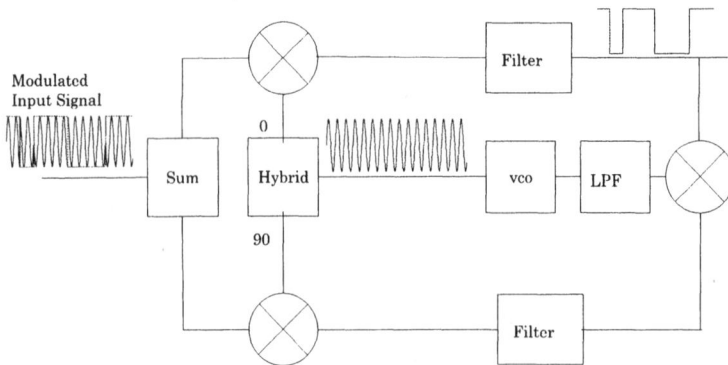

Figure 4-3 *The costas loop.*

The costas loop guarantees that the incoming signal is detected regardless of the phase of the incoming signal. If the signal is assumed to be a cosine wave without any phase difference in the demodulating cosine wave, then all of the signal will be present in Path 1. If the incoming signal is 90 degrees out of phase with the demodulating cosine wave, then the entire signal is present in Path 2. If it is somewhere in between, the signal will be detected in both paths, and the amplitude will be dependent on whether the cosine wave or the sine wave of the demodulating waveform is closest in phase to the incoming signal.

Orthogonal Polarization

Polarization techniques can be used when the polarization is 90 degrees from each other (vertical vs. horizontal). Another polarization scheme that uses orthogonality is LHCP/RHCP.

Polarization has been used for years to prevent unwanted signals from jamming the desired signals. If we assume that most of the unwanted signals are vertically polarized, then the system is set up for horizontal polarization which is a effective anti-jam system. Polarization can be used for other means. Separate signals can be sent out in orthogonal polarizations, which will be detected as separate channels or signals. Polarization is used in satellite communications where two signals are sent out of the satellite antennas using orthogonal polarization of the two signals. This is especially effective when small, cost-effective antennas are used on the ground with wide beamwidths that could cause interference to adjacent channels unless polarization is used.

Orthogonal polarization also increases the number of channels available for higher speed data within a given bandwidth using parallel channels. In addition, it can also increase the number of users for a given bandwidth.

4.2   Orthogonalizer Techniques

Orthogonality can be used for many applications. All of the uses are based around the fact that signals can be separated if they are orthogonal. Using this basis, jammers can be mitigated. The idea is to separate jamming signals from a composite jamming and desired signal, and then using the jamming signal to cancel out the jammer in the composite signal leaving only the desired signal.

Gram-Schmidt Orthogonalizer (GSO)

The *Gram-Schmidt orthogonalizer* mitigates jamming signals by using orthogonal principles. The jamming signal is forced to be orthogonal to the desired signal to minimize the effects of the jammer. All signals can be represented by the sum of weighted orthonormal functions.

$$S_1(t) = aX_1(t) \qquad\qquad 4.13$$
$$S_2(t) = bX_1(t) + cX_2(t)$$

where $X_1(t)$, $X_2(t)$ ... are orthonormal functions and a, b, c... are the weighting coefficients.

The constant a is simply the magnitude of $S_1(t)$, since $S_1(t)$ is in the same direction as $X_1(t)$, and the magnitude of $X_1(t)$ is equal to unity by definition. To solve for b, the second equation is multiplied by $X_1(t)$ and integrated.

$$\int S_2(t)X_1(t)dt = \int bX_1(t)X_1(t)dt + \int cX_2(t)X_1(t)dt \qquad\qquad 4.14$$

The second term on the right side of Equation 4.14 is equal to zero (the inner product of two orthonormal functions = 0); the first term is equal to b (the inner product of a orthonormal function with itself = 1). Solving for b produces:

$$b = \int S_2(t)X_1(t)dt \qquad\qquad 4.15$$

which is the projection of $S_2(t)$ on $X_1(t)$.

The constant c is determined by the same procedure except that $X_2(t)$ is used in place of $X_1(t)$ when multiplying the second equation. Therefore,

$$c = \int S_2(t)X_2(t)dt \qquad\qquad 4.16$$

which is the projection of $S_2(t)$ on $X_2(t)$. Note that $X_2(t)$ is not generally in the direction of $S_2(t)$.

The first orthonormal basis function is therefore defined as

$$X_1(t) = S_1(t) / a \qquad\qquad 4.17$$

where a is the magnitude of $S_1(t)$.

The second orthonormal basis function is derived by subtracting the projection of $S_2(t)$ on $X_1(t)$ from the signal $S_2(t)$ and dividing by the total magnitude.

$$X_2(t) = \left[S_2(t) - bX_1(t) \right] / c \qquad\qquad 4.18$$

where c is the magnitude of the resultant vector.

$$c = \left| S2(t) - bX1(t) \right| \qquad\qquad 4.19$$

A phasor diagram showing the projections of the vectors is provided in Figure 4-4.

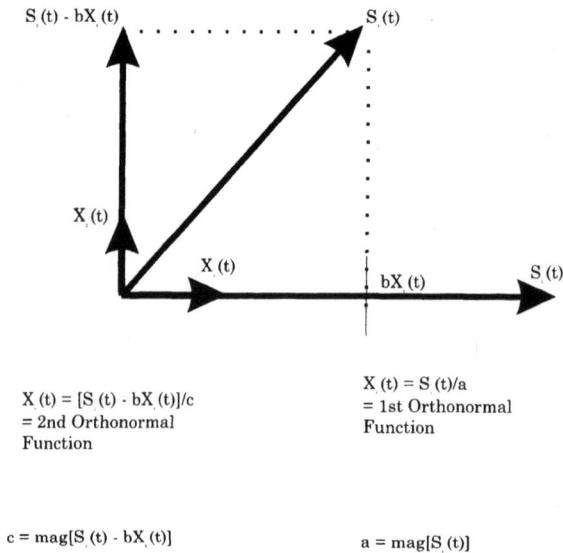

S (t) - bX (t)   S (t)

X (t)

X (t)   bX (t)   S (t)

X (t) = [S (t) - bX (t)]/c
= 2nd Orthonormal
Function

X (t) = S (t)/a
= 1st Orthonormal
Function

c = mag[S (t) - bX (t)]   a = mag[S (t)]

Figure 4-4 *Orthonormal vectors.*

As an example, Figure 4-5 shows a three-dimensional signal vector in the X/Y/Z plane using an X component with amplitude a, a Y component with amplitude b, and a Z component with amplitude c, with X , Y and Z all being orthogonal.

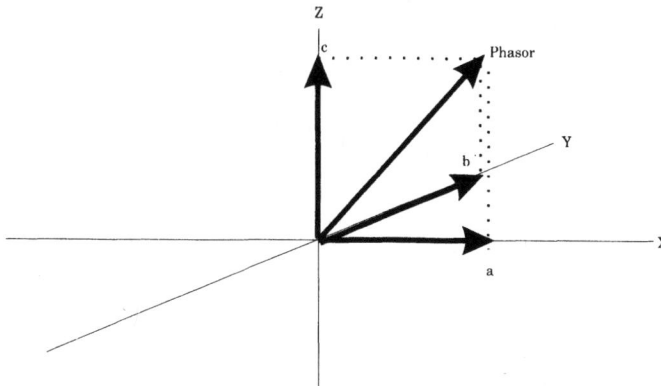

Figure 4-5 *A three-dimensional signal vector.*

The a is simply the magnitude in the X direction, b the magnitude of the signal in the Y direction, and c the magnitude of the signal in the Z direction.

Basic GSO

A basic GSO system is shown in Figure 4-6. The weight (ω_1) is chosen so that the two outputs, $S_1(t)$ and $S_p(t)$(desired signal) are orthogonal, which means that the inner product $< S_1(t), S_p(t)> = 0$. Suppose only a jammer exists in $S_1(t)$ and the signal plus jammer is equal to $S_2(t)$, as shown in Figure 4-6.

Taking the projection or inner product of $S_2(t)$ on the orthonormal function $X_1(t) = S_1(t) / |S_1|$ provides the amount of jammer present $S_2(t)$.

$$b = \langle S_2(t), \frac{S_1(t)}{|S_1(t)|} \rangle \qquad 4.20$$

This scalar quantity b multiplied by the orthonormal function $X_1(t)$ produces the jammer vector $bX_1(t)$. Subtracting the jammer vector from $S_2(t)$ gives the amount of signal present $(S_p(t)$ in $S_2(t)$ and eliminates the jammer.

$$S_p(t) = S_2(t) - bX_1(t) = \langle S_2(t), \frac{S_1(t)}{|S_1(t)|} \rangle \frac{S_1(t)}{|S_1(t)|} \qquad 4.21$$

In the implementation of these systems, the $|S_1|$ values are combined in a scale factor k where

$$k = 1/|S1(t)|2 \qquad 4.22$$

The final result is

$$Sp(t) = S_2(t) - k\langle S_2(t), S_1(t)\rangle S_1(t) \qquad 4.23$$

Therefore, the weight value is equal to

$$\omega = k <S_1, S_2> \qquad 4.24$$

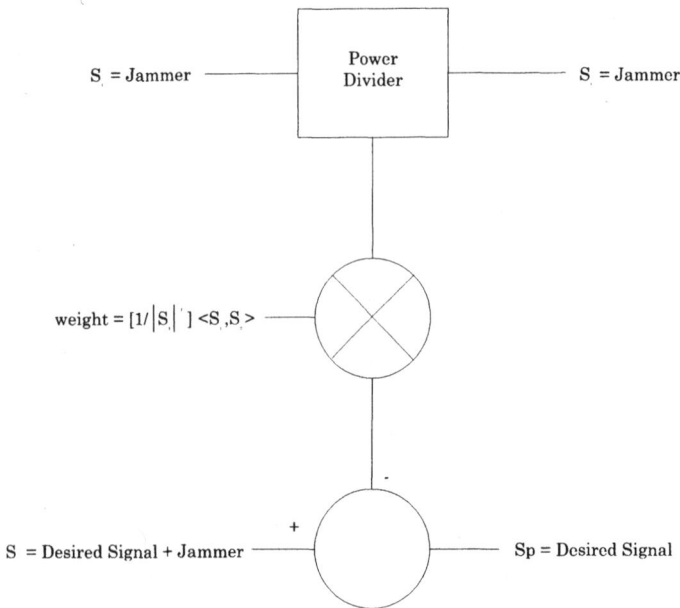

S = Jammer — Power Divider — S = Jammer

weight = [1/|S|] <S,S> —

S = Desired Signal + Jammer — Sp = Desired Signal

Figure 4-6 *Basic GSO system.*

4.3 Orthogonal Frequency Division Multiplexing (OFDM)

One of the more common techniques to provide higher data rates using parallel channels is *orthogonal frequency division multiplexing* (OFDM). OFDM uses multiple frequencies to send the data because each of these frequencies are orthogonal and independent detection of all the frequency channels used with minimum interference between channels is possible. OFDM is used extensively in both wireless applications and power line communications. It has been used in wireless LANs, cellular phones and other wireless devices. It is ideal for broadband communications and networking over the existing power lines in a home or small office.

OFDM is used in conjunction with a digital modulation scheme, such as BPSK, QPSK, and QAM. This allows higher throughput of the data utilizing multiple channels in parallel and allows for a very robust system to drop frequency channels that are being jammed or unusable due to the amplitude spectral response.

Basic orthogonality means that the inner product of two orthogonal frequency channels are integrated to zero over a specified integration time, and the inner product of the same frequency is equal to one.

$$\langle Xn(t), Xm(t) \rangle = \int_0^t Xn(t)Xm(t)dt = 0 \quad where\, n \neq m \qquad 4.25$$

$$\langle Xn(t), Xm(t) \rangle = \int_0^t Xn(t)Xm(t)dt = 1 \quad where\, n = m \qquad 4.26$$

Therefore, if the frequencies are orthogonal, their inner products are equal to zero. Taking the inner product of all the frequencies with the desired frequency, only the incoming desired frequency will be present.

One way to obtain orthogonal frequencies is to ensure that the frequencies are harmonics of the fundamental frequency. Then, by choosing the integration time, the inner product will be zero. An example of orthogonal frequencies is shown in Figure 4-7. This example contains the fundamental, the second harmonic, and the third harmonic.

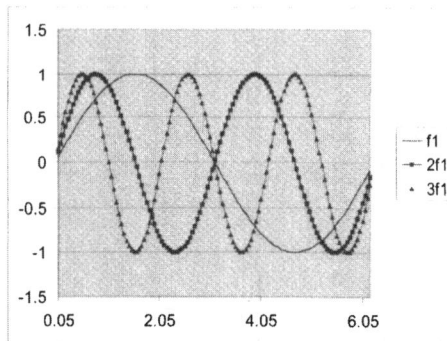

Figure 4-7 *Orthogonal frequencies producing an inner product of zero.*

Any combination of these frequencies is multiplied and integrated with the limits shown in Figure 4-7, and the result is equal to zero. If the inner product of any one of these frequencies with itself is used, then the inner product is not zero, and the specified signal is detected.

The example above can be visualized by multiplying every point on the curves and sum the values. The values will be equal to zero for the orthogonal frequencies and equal to a one for the same frequencies.

Using orthogonal frequencies reduces adjacent channel interference and allows for more channels to be used is a smaller bandwidth (see Figure 4-8).

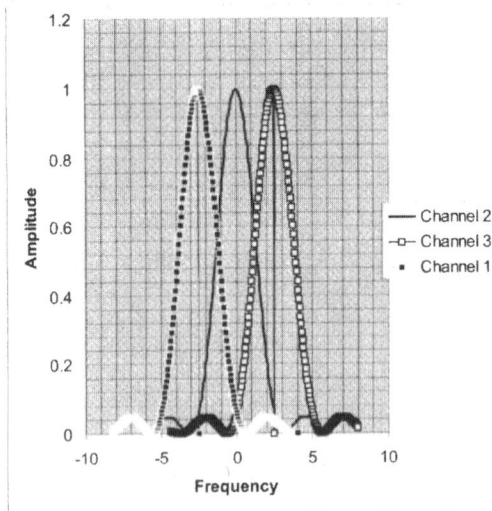

Figure 4-8 OFDM allows frequency channels to overlap with minimal adjacent channel interference.

Coded OFDM

Another version of the standard OFDM is coded OFDM (COFDM). This scheme uses several techniques to improve on the standard OFDM and developed methods to reduce the effects caused by multipath. COFDM uses *instantaneous fast fourier transforms* (IFFT) modulation, fixed bit loading, and sophisticated coding schemes to overcome the fades due to multipath. The COFDM standard that has been set for digital video broadcasting is OFDM2.

VOFDM

Vector OFDM (VOFDM) is a type of OFDM that was developed by Cisco Systems. This technology is supported by an industry coalition. The VOFDM system uses *time division multiplexing* (TDM) and packet communications. VOFDM is designed for wireless broadband networks using two different frequency bands; 5 GHz, and 28 GHz (which is used for LMDS). The MAC layer will be similar to the Ethernet packet standards.

4.4   Summary

Orthogonal signals are used extensively in communications because they can be received and demodulated as separate data streams with very little interference between the orthogonal signals. Quadrature demodulation is used to eliminate the ambiguity of the phase of the incoming signal, but also is used to provide a means of either sending more data or providing two separate data streams that can be demodulated separately. Antenna polarization is used to allow two channels to operate on the same frequency in quadrature. This can be either horizontal/vertical or LHCP/RHCP.

GSOs can be used to reduce the effects of jamming signals. The jammer and the signal are forced to be orthogonal so that the jammer can be eliminated and the signal be detected. GSOs are based on the inner product of orthogonal signals being equal to zero. Therefore, if they are orthogonal and the inner product is zero, then the desired signal can be easily detected in the presence of an unwanted signal or jammer. This approach assumes that the jammer signal level is much higher than the desired signal level. The basic GSO uses two inputs: one of the inputs contains more signal than jammer. This applies to having two antennas with one directed towards the signal providing higher power. The error signal for feedback in updating the weight value is produced by subtracting the weighted reference input signal from the received signal which contains the higher level of desired signal. When the weight has converged, the jamming signal is suppressed.

OFDM is a method used in many communication applications where high-speed data or multiple users are desired. OFDM provides an optimal, spectrally efficient system for use with multiple users or the combination of parallel frequency channels for overall high data rates used for broadband communications and home networking.

4.5   References

1. Bullock, Scott R. *Transceiver System Design for Digital Communications*. Atlanta: Noble Publishing, 1995.

2. Inglis, Andrew F. *Electronic Communications Handbook*. New York: McGraw-Hill, 1988.

3. Crane, Robert K. "IEEE Transactions on Communications." vol. Com-28, no. 9 (September 1980).

4. Schwartz, Mischa. *Information, Transmission. Modulation and Noise.* New York: McGraw-Hill, 1980.

5. Haykin, Simon. *Communications Systems*. New York: John Wiley & Sons, 1983.

6. Holmes, Jack K. *Coherent Spread Spectrum Systems*. New York: John Wiley & Sons, 1982.

7. Bullock, Scott R. "Use Geometry to Analyze Multipath Signals." *Microwaves & RF* (July 1993).

8. Bao-yen Tsui, James. *Microwave Receivers with Electronic Warfare Applications*. New York: John Wiley & Sons, 1986.

9. Bullock, Scott R. "Phase-Shift Keying Serves Direct-Sequence Applications." *Microwaves and RF* (December 1993).

10. Bullock, Scott R. "High Frequency Adaptive Filter." *Microwave Journal* (September 1990).

11. "Satellite Communications Handbook, Fixed Satellite Service," *International Telecommunications Union, International Radio Consultative Committee* (Geneva, 1988).

5

Networking for Home and Small Office

etworking devices in the home and *small office/home office* (SOHO) that provide one complete network that can be controlled from a central location have become increasingly popular. Many companies are providing the link from the providers to the user using twisted pair wire, coax cable, fiber optics, wireless and satellite with the objective to increase the bandwidth for both high-speed connections and provide this bandwidth to an increasing number of users.

Networks have been in use for some time; however in the recent years and with the increase in the number of users and computers, the Internet, and all of the different ways that information is being brought to the home, networking and distribution are in high demand. The technologies that are under consideration for home distribution include wireless, power line, and phone line communications. Wireless will be the most successful technology because it affords networking in the home without requiring wires and a new infrastructure.

5.1   Local Area Networks (LAN)

Various types of networking, protocols, and hardware devices have been developed in the past with various standards describing these networks to provide interoperability with other devices. When a network is used in a local area, for example the SOHO, it is referred to as a *local area network* (LAN), versus a *wide area network* (WAN), which covers and networks a wider space. A LAN links multiple devices for data transfer and sharing of resources in a small geographical area, such as the home or SOHO. Several PCs can be connected as a LAN so that file and print sharing, software, memory sharing, and games can be utilized as one system. A server can be used in a LAN to provide extra storage of data, software, as well as to operate and control the system.

The LAN topologies describe how the network is connected to each of the users of the network. The main three types of networks include the star, bus, and ring, with variations and branching of the main system.

Star Topology

The star topology contains a central node that all other nodes are connected to, as shown in Figure 5-1. Therefore, none of the other nodes are connected to each other directly, however, direct communications between nodes, or node to all of the other nodes is possible, but must travel through the central node first.

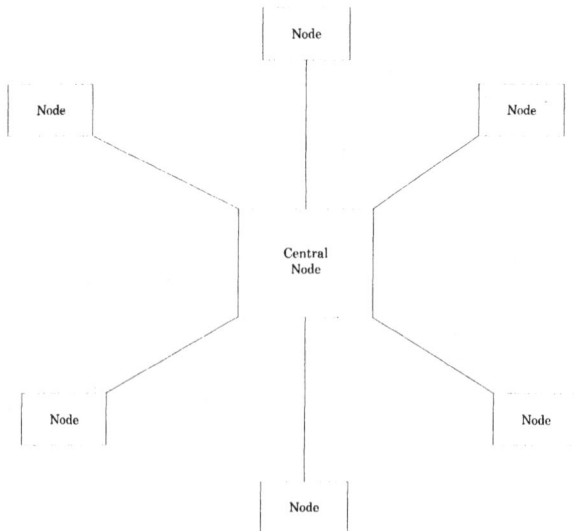

Figure 5-1 *Star topology for home networking.*

The central node can be passive or active. If it is passive, it provides coupling to the other nodes. If it is active, it acts as a repeater between the nodes. The star configuration is used when the central node is used extensively by the other nodes, for example, if it contains the storage for backing up all of the other nodes. The star topology is easier to troubleshoot if the system becomes inoperable. One of the drawbacks of the star topology is that the reliability of the system is dependent on the reliability of the central node.

Bus Topology

The bus is a communications channel in which all of the nodes are connected to a bus, or wire, and each of the nodes or devices communicate through the bus, as shown in Figure 5-2.

Bus Connection

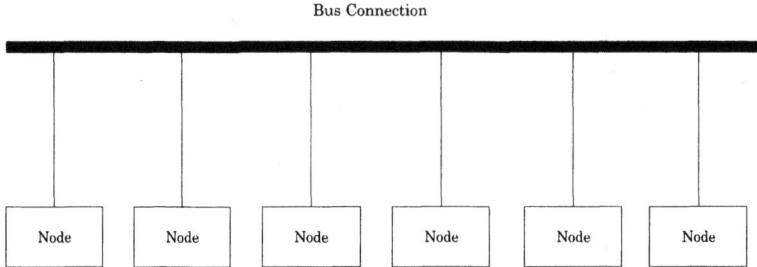

Figure 5-2 Bus topology for home networking.

Unlike the star configuration where there is one central node, the bus configuration connects all of the nodes to the bus. Therefore, a node can communicate with any other node or all the nodes via the communication bus (see Figure 5-2). In this type of topology, each node can act as a controller. Therefore, each node has the protocol for that particular network. The node listens to see if the bus is busy, and if clear, instigates a call to another node/nodes for communications. The other nodes are listening and respond if they are requested to do so. Various techniques and protocols prevent collisions and multiple nodes requesting the bus at the same time. The bus topology is very reliable since the communications link only depends on the communicating active nodes and the bus, which is generally higher than most topologies.

Ring Topology

Another method of connecting a network is a ring or loop. The ring topology connects all of the nodes in a daisy chain configuration (see Figure 5-3). The advantage of the ring configuration is that the transmitted signal is sent around the ring and back to the sender where the sender can do a comparison and perform error detection. The disadvantage is that since there could be a problem with any node, it may be difficult to determine exactly where the fault occurred. Any node could break the chain or the communication channel, so the reliability of the system is equal to the reliability of all the nodes being fully functional.

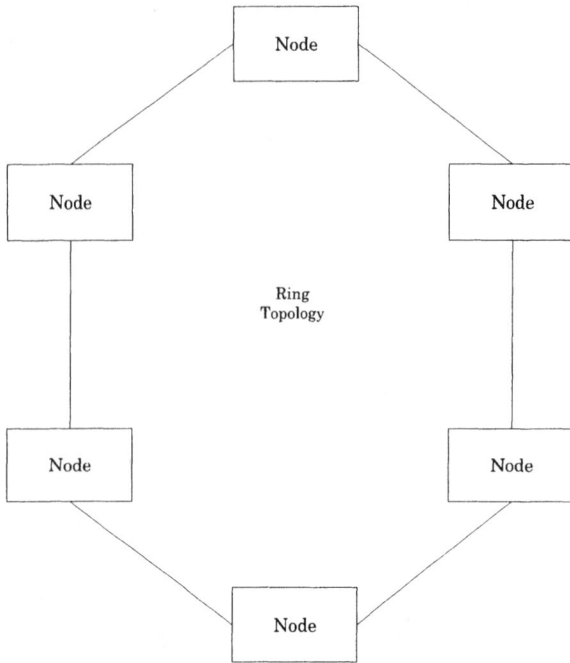

Figure 5-3 Ring topology for home networking.

5.2 LAN Standards

LAN Standards have been established to set the direction and interoperability of different manufacturer's hardware. The IEEE established a local area network standards committee 802 which adopted the 802.3 standard for bus topologies using carrier sense, multiple access with collision detection CSMA/CD and the 802.5 standard dealing with token rings.

Ethernet

Ethernet uses a CSMA/CD scheme, which relies on a node that senses if the bus is clear and then sends out a transmission and monitors if there is another node trying to transmit. If the node detects another transmission, then it sends out a signal telling all nodes to cease transmission. The other nodes cease for a random period of time and then try again if the channel is clear. This helps to prevent an endless loop of collisions between nodes trying to use the bus. Propagation delay can also increase the probability of collision. Someone may be using the system, or the other node may not have received his transmission due to the propagation delay so the node attempts to use the system.

There are four common standards using Ethernet: 10BASE-5, 10BASE-2, 10BASE-T, and 100BASE-T. The designation tells the user the physical characteristics. The 10 describes the operating speeds, 10 Mbps, the BASE stands for baseband, the 5 is distance (500 meters without a repeater) and the T stands for unshielded twisted pair. 100 BASE-T provides 100 Mbps data rates under the IEEE 802.3u standard using CSMA/CD technology. The 100BASE-TX is a version that provides 100 Mbps over two pair of CAT-5 unshielded twisted pair (UTP) or two pair of Type 1 shielded twisted pair (STP).

Token Ring

The token ring protocol prevents data collisions by allowing only the node that has the ring to transmit. Therefore, by definition, only one node can transmit if it has the ring. One of the nodes is designated the active node and sends a packet of data, the token, around the ring. As each node receives the token, it checks for the address and if it has the correct address, it receives the message, appends its message to it and passes the token with the message on to the next node. The ring topology prevents multiple transmissions, but is a slower method of sending data since each node has to wait for the token before it can transmit.

5.3   Fiber-Distributed Data Interface

Fiber optics are utilized to increase the bandwidth and data speeds that are required to transport voice, data, and video for home/SOHO applications. The *fiber-distributed data interface* (FDDI) was established to utilize the ring topology. It increased the operating speed to 100 Mbps and ranges to 2 km.

5.4   Universal Serial Bus (USB)

The *universal serial bus* (USB) is an external bus standard providing data rates up to 12 Mbps. The new standard USB2 supports up to 480 Mbps. A USB can daisychain up to 127 peripherals to a single USB port. This is convenient for the end user because there is virtually no problem concerning the availability of the connector and number of the devices connected to the USB.

The USB provides a plug-n-play system, which is a tremendous advantage to the end user. The subscriber simply plugs in the device. The computer detects the connection, loads in the drivers or asks for the necessary software, which is provided by inserting the disk, and then follows instructions. Therefore, the system is automated and the chain of events begins as soon as something is connected to the USB.

Another advantage to using the USB is that it can supply power to devices that are plugged into the USB port. Thus, the device does not need to be plugged into anything else. However, the power output is limited and may not be adequate for many devices.

Since the USB is truly a plug-n-play system, the end user is not required to open up the computer and install personal computer cards, change the dip switch locations, allocate memory, or choose the interrupts which is necessary for other type of networking systems. The USB port will eventually replace the serial and parallel ports of the computer.

A third advantage for using USB is that USB devices can be added, changed, or disconnected without turning off the computer. This really saves time since there is no waiting as the computer is booting up in the process.

There are many applications for using USB. The USB can provide a plug-n-play networking system connecting multiple devices and peripheral providing communications with several devices.

5.5   Cable Networks

Coaxial cables have been used mainly for delivering television to the home. In recent years, the cable services are providing data communications, including the Internet and telephone services. *Data over cable service interface specifications* (DOCSIS) help to provide interoperability between cable devices from multiple sources and manufacturers.

The cable connections can provide the "last mile" connection. Cable can also provide distribution and networking for neighborhoods, multiple houses, and small and large businesses.

5.6   Network Layers

The standard for network layers is the *open system interconnection* (OSI) seven-layer model. Some of these layers may not be included in the design of a network system, but the following list gives an overview of the different types of layers and details what they do. More detailed description can be found in the references to this chapter. The different layers are shown in Figure 5-4.

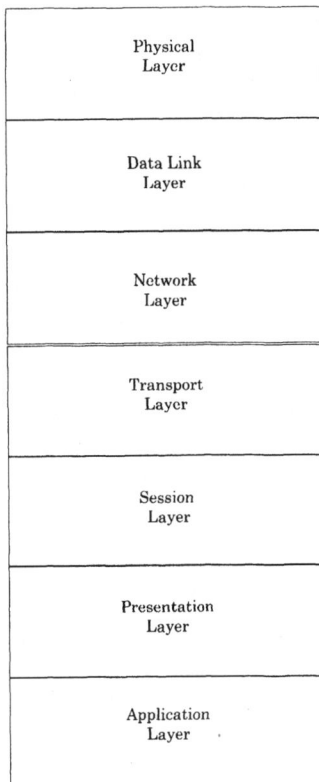

Figure 5-4 *OSI seven network layer model.*

1. **Physical layer.** This is the lowest level network layer. It is mainly concerned with voltages, currents, impedances, power, signal waveforms, but also with connections or wiring, such as the RS-232 serial interface standard.

2. **Data link layer.** This layer is used to communicate between the primary and secondary nodes of a network. It is involved with activating, maintaining, and deactivating the data link. It is also concerned with framing the data and error correction and detection methods.

3. **Network layer.** This layer specifies the network configuration and defines the mechanism where messages are divided into data packets.

4. **Transport layer.** This layer controls end-to-end integrity, interface or dividing line between the technological aspects and the applications aspects.

5. ***Session layer.*** This layer controls network availability, log-on, log-off, access, buffer storage and determines the type of dialog, simplex, duplex.

6. ***Presentation layer.*** This layer specifies coding, encryption, compression.

7. ***Application layer.*** This layer communicates with the users program and controls the sequence of activities, sequence of events, general manager.

5.6   Summary

Many types of topologies can be used to connect several devices together. The star topology utilizes a central node that connects to all the other nodes. The star topology is easy to troubleshoot because the central node can be used as a storage for all of the other nodes. The disadvantage is that if the central node has a problem, the whole system has a problem. In a bus topology, all of the nodes are connected to a bus. Any node can talk to any other node directly. The bus topology requires a protocol to prevent collisions since the nodes do not all go through a single node. The ring topology provides a means for error detection because the signal goes around the ring to ensure that the data was received properly. The disadvantage to the ring is troubleshooting, since any node could cause the whole system to be disabled.

Standards have been developed and continue to be developed for LANs. The IEEE has set standards labeled 802.xx for networking. Some of the standards relate to the Ethernet which uses a CSMA/CD scheme to prevent collisions.

Other networking technologies used include fiber, cable, and USB. The USB is a true plug-n-play method; it of connects multiple devices to a network with minimum user interface. The OSI 7-layer model provides a way for interoperability between products and to establish a way to define the portions of the design.

5.7   References

1. Fike, John L., George E. Friend, Stephen J. Bigelow. *Understanding Telephone Electronics*. Indianapolis: SAMS Publishing, 1996.

2. Bullock, Scott R. *Transceiver System Design for Digital Communications*. Atlanta: Noble Publishing, 1995.

3. Tomasi, Wayne. *Advanced Electronic Communications Systems*. Fourth ed. Englewood Cliffs, NJ: Prentice Hall, 1998.

4. Papi, Zdzisla and Andrew Simmonds. "Competing for Throughput in the Local Loop." *IEEE Communications Magazine* (May 1999).

5. Lough, Daniel L., T. Keith Blankenship and Kevin J. Krizman. "A Short Tutorial on Wireless LANs and IEEE 802.11." (March 2000).

6

Power Line Communications

A *power line communications* (PLC) system modulates voice, data, and video signals and sends them over the existing A/C power lines to the h ome. Power lines offer the best existing infrastructure and the lowest cost broadband communications and home networking presently available. The signals are simply A/C coupled to the power line through a transformer or balun. The frequencies are low enough to minimize radiation as the signal travels down the power lines to the receiver. The range is adequate to service nearly all of the outlets. Even with connections on different phases, communications is possible though coupling between the phases (*phase jumping*). Communications through the power lines are classified as current carrier devices and provide an excellent way to establish networks and communications in an already existing infrastructure.

As is the case with all systems, noise and jamming signals are also present on the power lines. Also, multipath and impedance mismatches occur, which can cause some of the band to be unusable. Because a power line is hard wired and does not rely on electromagnetic waves through the air, signal blockage is irrelevant. However, power lines have to contend with appliances and other devices plugged into the outlets which can cause attenuation, mismatches and noise.

PLC devices that are presently in use have operated with minimal dropouts and have been very reliable in spite of the potential problems that exist on the power lines. Most of the newer systems today utilize spread spectrum, parallel channels, or frequency selection techniques, or a combination of them to help mitigate the noise and jammers and overcome the attenuation and impedance variations that are present on the power lines.

In addition, these techniques are used to increase the data rates that are capable on the power lines. Some PLC systems are capable of 11 Mbps, which corresponds to the Ethernet type speeds for networking. Thus, the power line medium is a viable solution for low-cost, high-speed broadband communications and home networking system.

6.1   Communication Over the Power Lines

Existing power lines in a home can be used as a medium for communications and distribution of voice, data, and video signals. Instead of installing new wires, this technology uses the existing wiring in the home, the A/C power wiring. By sending out modulated information over relative low frequencies on the power line, the signals are transmitted along the lines instead of radiating through the air (see Figure 6-1). Multiple devices can be connected to the telephone line, including the standard telephone, fax machines, answering machines, cable set top boxes and satellite TV that require a telephone company (telco) return path, modems and connection to the Internet, and many other uses (see Figure 6-1). Multiple devices can be connected to the same system, however, without networking the devices, multiple lines or multiple access channels, only one communication device can operate at a time.

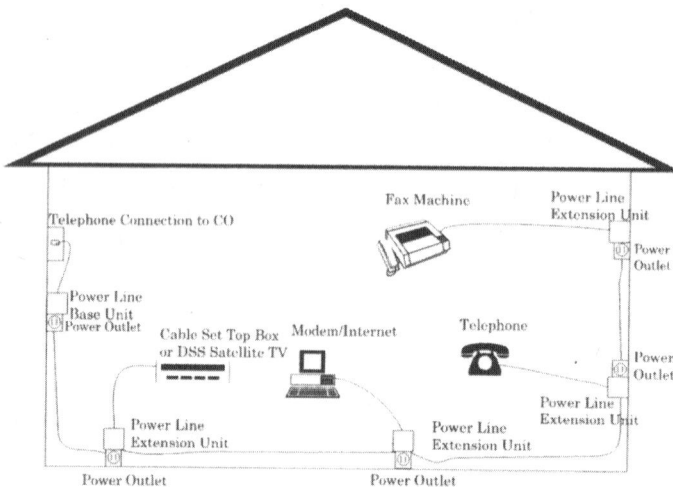

Figure 6-1 *Communication over the power lines.*

Therefore, every outlet in the home can be used for communications and networking. According to the Part 15 Rules and Regulations, this method is called *carrier current* and is classified as an unintentional radiator. Using this method eliminates the upconverter and the antenna.

The power line infrastructure can be used to network the home to be used for print sharing, file sharing, games, Internet access, telephone, *voice over Internet protocol* (VoIP), music, and video (see Figure 6-2). The home network is focused on providing networking without running wires. This provides easy installation and an immediate plug-n-play system. The network connections are called nodes; the system resembles a standard Ethernet-wired system. Broadband communications and home networking over the power line follows a similar path as other technologies in providing and distributing high-speed information throughout the home.

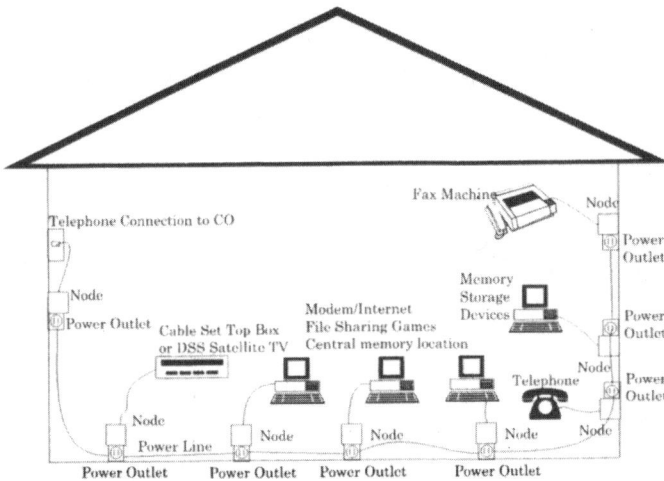

Figure 6-2 *Power line networking of multiple devices.*

Some attempts have been made to use the power lines to distribute signals from the power stations to the home (see Figure 6-3). However, most of these attempts have not proven effective. Problems occurred with respect to power output levels, mismatches and stubs that act as antennas and radiate the signals, distribution boxes and power meters that attenuate the signals. Also, the amount of power output required was generally too high to be practical. Given that the wiring is in place to service homes and offices that are all connected to the power lines, more attempts to use power lines for distribution of communications and networking can be expected.

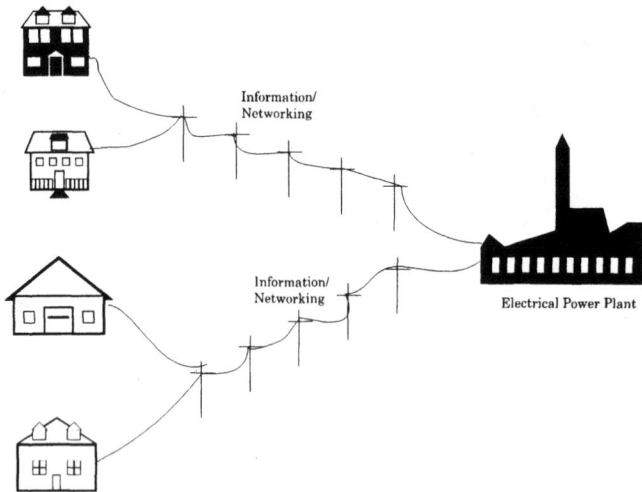

Figure 6-3 *PLC from power station to the home.*

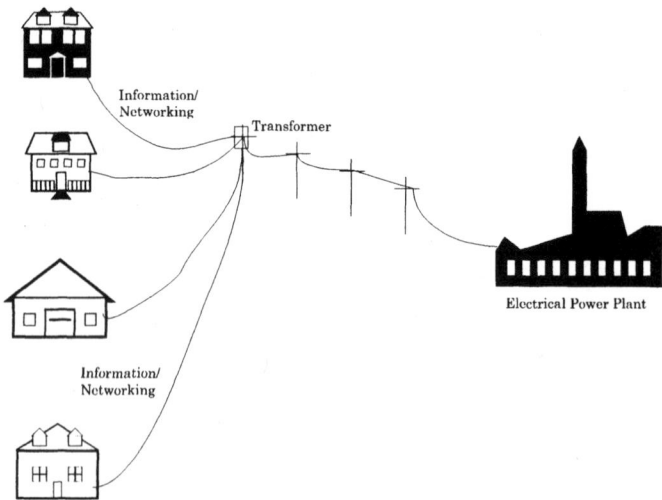

Figure 6-4 *PLC distribution from a power transformer central point.*

Another application of PLC outside the home environment distributes communications from a power transformer to several homes that are tied to this central point (see Figure 6-4). This intermediate point does not require as high of power output. The distance traveled is much shorter, which makes it more feasible to overcome the problems. The information is sent to the transformer point either via a hard-wired cable, fiber optics, or RF or microwaves.

6.2   Power Line Modulation

Several frequency bands are used to send and receive signals via the power line. One band that is popular for lighting control, appliance control, and other applications is in the 100 to 500 kHz band. This low frequency propagates well on the power line. The different stub lengths are generally too short to cause much problem with reflections and radiation. However, power in the US comes in as a 220 VAC three-phase signal. The phases are split into two 110 VAC phases and isolated. This frequency band does not "jump" phases very good. A means of A/C coupling these phases together by a capacitor is generally required to ensure operation on different phases (see Figure 6-5). Note that if the frequency were higher, the coupling between phases would be adequate without adding the A/C coupling capacitor.

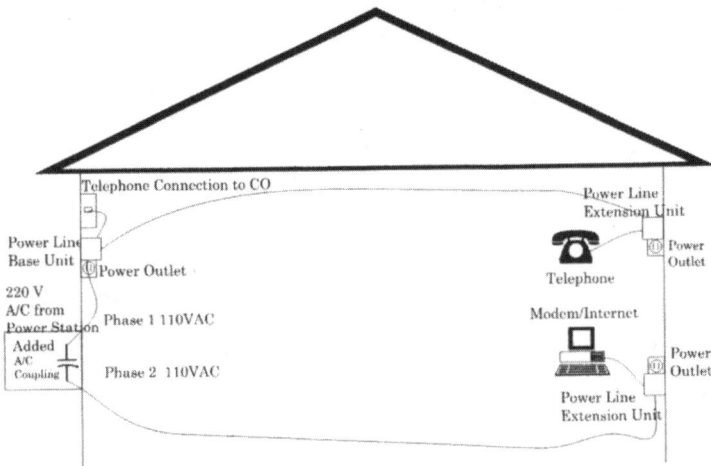

Figure 6-5 Different phases require coupling for low frequency PLC.

The 1 to 30 MHz band is predominantly used for broadband communication applications and wireless voice, data, music, and video applications. It provides higher data throughput, better phase-to-phase communications, phase jumping, and is the preferred band for high-speed power line communications. Coupling the phases together is generally not necessary for this frequency band; however, coupling would reduce attenuation and improve performance. The 1 to 30 MHz frequency band, along with high-level modulation schemes and parallel techniques, allows higher data rates to be transferred through the power lines.

PLC communications is considered an unintentional radiator and is generally confined to the power line. However, since this band is shared among many other users, frequency coordination and frequency selection is important to minimize the interference between users.

Various types of modulation schemes can be used to send the data over the power lines, ranging from *frequency modulation* (FM) methods — used to provide connections for telephones, faxes, and modems and a return path to the telephone line for cable *TV set top boxes* (STB) and *direct satellite service* (DSS) — to digital modulation techniques, including BPSK, QPSK, QAM, FSK and FH. To increase the data throughput, these modulation formats include parallel means to send data that, when combined, produce an overall high-speed data rate. One of the technologies that is commonly used for PLC is OFDM. OFDM uses parallel overlapping orthogonal frequency channels to provide high-speed data rates with efficient use of the bandwidth. OFDM is also used to select the bands to be used and to eliminate bands that may cause interference to other users in the band.

Error detection and correction schemes are designed into the systems to improve the communications over the somewhat noisy medium and inherent jamming signals. Spread spectrum techniques may be applied to further enhance the reliability for these power line products.

Standards make the products that communicate over the power lines interoperable and prevent multiple power line carrier products with different standards from causing interference to each other.

6.3   The Power Line Infrastructure

Power line communications appears to be the simplest and cheapest solution to communicate using voice, data, music, video, Internet access, and networking. Can the power line communications medium be reliable and provide the wide bandwidths that are needed to satisfy the continually increasing demands for higher data rates and rising number of users? Advanced technology needs to be designed to overcome the noise that is present on the power line and the impedance mismatches caused by the different lengths of power lines and different terminations which cause peaks and nulls in the amplitude response of the spectrum (see Figure 6-6).

Currently, frequency agile systems, frequency smart systems, power control and adaptive systems are being considered to improve the performance over the power line. With continual improvements in technology, the power line infrastructure will provide the needed distribution of signals in the home.

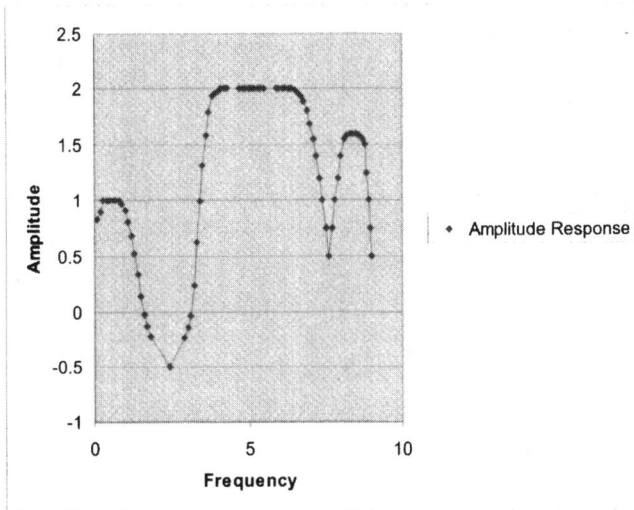

Figure 6-6 *Amplitude response of the spectrum in a power line.*

6.4 Advantages and Disadvantages Using Power Line Communications

The advantages for using the A/C power lines for communicating signals in the home are

1. existing infrastructure with extremely thorough coverage of the home or small office,

2. low cost,

3. no signal blockage,

4. no RF multipath because the signal is confined to the power line.

The disadvantages for using the A/C power lines for communicating signals in the home are

1. attenuation between different phases of the A/C power line,

2. power lines are subjected to noise and jamming signals from multiple devices that are plugged into the outlets,

3. mismatches and reflections cause nulls in the frequency spectrum,

4. power line loads and noise levels are dynamic, depending on what is connected (plugged in) to the power lines.

Despite these disadvantages, millions of PLC devices have been installed worldwide with tremendous success. Many of these disadvantages for power line communications can be overcome by more sophisticated and advanced design techniques.

6.5 Summary

Power line communications is a viable medium for sending and receiving information, which includes voice, data, music, video, Internet access, networking, and many other applications. PLC uses an already existing infrastructure, the A/C power lines, for distribution of these types of signals without adding new wiring to the home or small office. Because of the existing infrastructure and lack of high frequency RF components or antennas, this method for sending and receiving information appears offer the lowest cost and the best coverage at the present time. Nearly all types of modulation schemes can be used for sending and receiving signals over the power line; however, for high speed, reliability, and security, digital modulation is the preferred method. These digital modulation schemes along with parallel OFDM type techniques provide the demands for broadband communications and home and small office networking.

Standards regarding the interoperability of multiple devices are in place to prevent interference and jamming between users. These standards are being modified and upgraded as the technology and higher data rates are being incorporated. Different frequency bands are being used for different purposes; spread spectrum and adaptive systems are being designed in order to co-exist and for interoperability. The advantages of a power line communication system far outweigh the disadvantages.

6.6 References

1. Bullock, Scott R. *Transceiver System Design for Digital Communications.* Atlanta: Noble Publishing, 1995.

2. Inglis, Andrew F. *Electronic Communications Handbook,* New York: McGraw-Hill, 1988.

3   Holmes, Jack K. *Coherent Spread Spectrum Systems*. New York: John Wiley & Sons, 1982.

4. Bullock, Scott R. "Use Geometry to Analyze Multipath Signals." *Microwaves & RF* (July 1993).

5. Bullock, Scott R. "Phase-Shift Keying Serves Direct-Sequence Applications." *Microwaves and RF* (December 1993).

7

Telephone Line Communications

Telephone lines provide an already existing infrastructure in the home that can be utilized for broadband communications and home networking. A communications node can be connected to every telephone outlet in the house to provide a network through the telephone lines. However, most homes are not wired adequately to provide ubiquitous connections in the home. Some homes may only have one telephone outlet in the entire house. Therefore, the telephone line networking system is limited in coverage.

One of the ways to increase the coverage and still use the telephone lines for networking is to combine the telephone network with either power line communications, wireless communications, or all three methods. This hybrid provides excellent coverage throughout the home and small office.

7.1 Communications Over the Telephone Line

Telephone line communication transmits signals over the existing telephone line infrastructure (see Figure 7-1). Therefore, wherever there is a telephone line present, the voice or data will be transmitted without running additional wires. Communication over the existing telephone line is used for simultaneous shared Internet access, printer and file sharing, and network gaming.

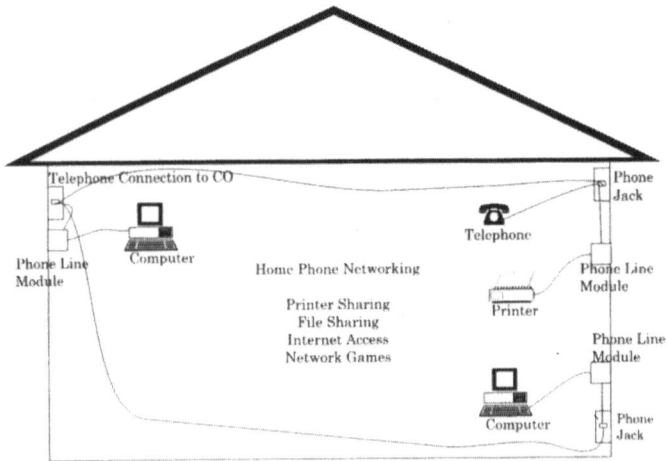

Figure 7-1 *Communication using existing telephone lines*

7.2 Telephone Line Modulation

The data and voice is coupled to the telephone line and transmitted to the other devices via the telephone lines. Telephone line communications uses an Ethernet network and the telephone lines for the infrastructure. Telephone line developers plan to provide USB and Ethernet adapters to the home network.

Since the telephone line is used for both voice and data modem applications, the networking applications cannot interfere with these existing signals in order to be in compliance with Part 68 of the FCC rules and regulations. This is accomplished by using FDM. Since the voice and data frequency bands are below about 1 MHz, the home networking signals operate above 2 MHz, thus avoiding interference with the other signals on the telephone lines. Therefore, the end user can operate the telephone devices and high-speed DSL modems for the Internet at the same time that the home networking is operating with minimal interference.

The 1.0 standard operates at 1 Mbps with a 5.5 to 9.5 MHz carrier frequency using the IEEE 802.3 standard, which is similar to Ethernet. The next generation is targeting 10 Mbps data rates.

7.3 Telephone Line Standards

The standard for telephone line communications was set by the *home phone network alliance* (Home PNA). Home PNA has two standards, the

1.0, and the 2.0. The 1.0 standard is capable of 1 Mbps and was released in the fall of 1998. The 2.0 standard has increased this speed to 10 Mbps and requires that this be backwards compatible with the 1 Mbps 1.0 standard.

7.4   The Telephone Line Infrastructure

This telephone data transfer system uses the telephone lines that already exist in the home. This technique modulates the voice, data, music, and video over the existing infrastructure in the home (see Figure 7-2).

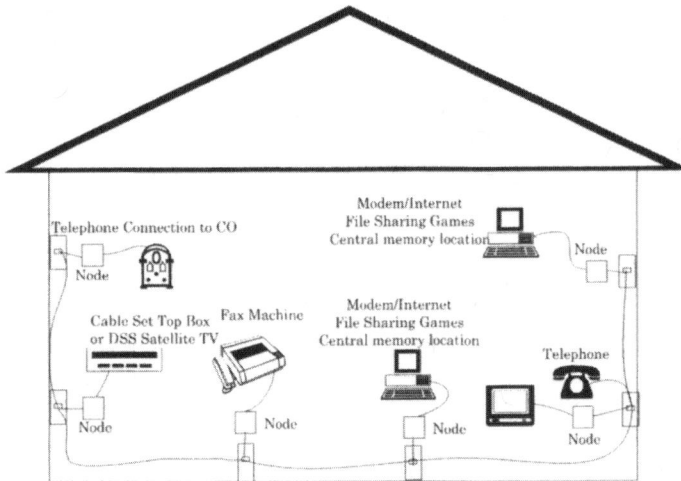

Figure 7-2 *Telephone line used to connect multiple devices in a home.*

This system is ideal if the existing home has multiple telephone lines and has telephone lines running in nearly every room in the house. In many homes, however, the telephone line infrastructure does not provide a means to distribute signals into every room in the house. This is a major limitation to Home PNA. The power line and RF methods provide a much better solution for full coverage throughout the home.

7.5   Advantages and Disadvantages Using the Telephone Lines

The advantages for using the telephone lines for communicating signals in the home are

1. existing infrastructure with good coverage in some of the newer homes and good coverage in an office environment,

2. low cost,

3. no signal blockage,

4. no RF multipath because the signal is confined to the telephone line,

5. not as susceptible to noise, changing conditions and variations in impedance.

The disadvantages for using the telephone lines for communicating signals in the home are

1. mismatches and reflections cause nulls in the frequency spectrum,

2. limited coverage in most of the homes, especially in the older homes,

3. additional infrastructure for better coverage will add additional costs.

The nulls in the frequency spectrum can be overcome by more sophisticated and advanced design techniques; however, the limited infrastructure will cause problems.

A combination with other technologies, either power line or RF may be another viable solution to increase coverage. The combination of telephone line and power line distribution is a viable option, since the telephone line infrastructure is not as ubiquitous as the power line. Combining the telephone line system with the power line system provides nearly 100 percent coverage in the SOHO market (see Figure 7-3).

The combination of the telephone line system with both the power line and RF wireless systems provides a complete system with coverage throughout the house. Many applications require wireless technology and more and more future products demand the versatility of mobility. The complete solution using phone lines, power lines, and RF provides the ultimate solution to broadband communications and home networking (see Figure 7-4).

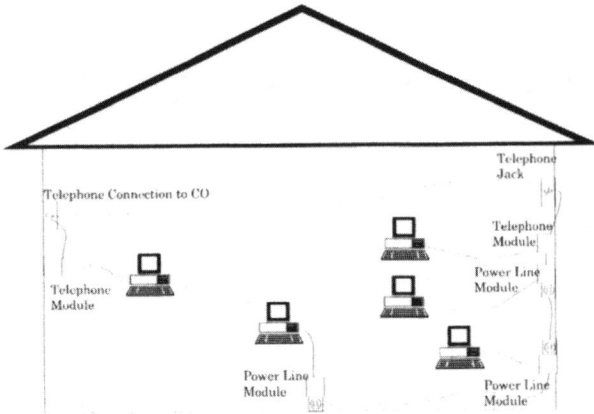

Figure 7-3 *Combination of telephone and power line systems.*

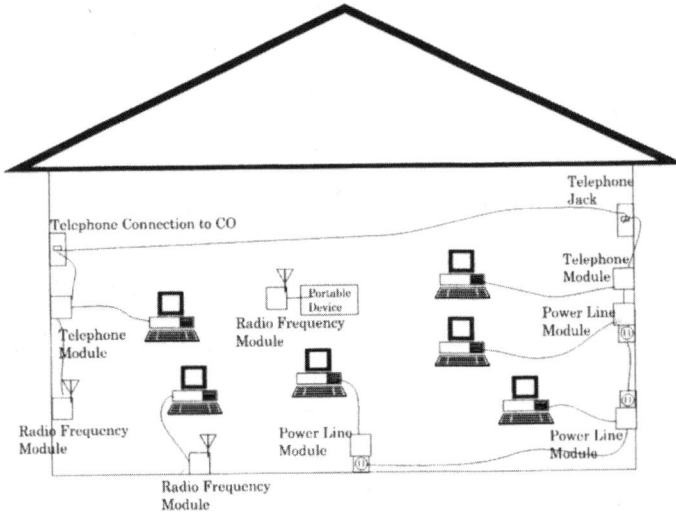

Figure 7-4 *Complete system using telephone lines, power lines, and RF.*

7.6 Summary

Telephone line communications is an adequate medium for sending and receiving information, such as voice, data, music, video, Internet access, and networking devices. This system uses an already existing infrastructure. In a home where there is adequate infrastructure or in an office environment, the distribution of these types of signals can be viable without adding new wiring to the home or small office. Because of lack of high-frequency RF components or antennas, this method for sending and receiving information offers a low-cost solution.

Nearly all types of modulation schemes can be used for sending and receiving signals over the telephone line; however, for high speed, reliability, and security, digital modulation is the preferred method.

The Home PNA group has established two standards, 1.0, and 2.0. The newer standard, 2.0, provides high-speed data rates, 10 Mbps, for use in high speed, broadband communications and home and small office networking. The standards will be upgraded as the technology and higher data rates are being incorporated. The major disadvantage concerning the telephone line communications system is the lack of coverage in the average home using the existing infrastructure. Other methods or additional wiring is needed for full coverage in the home.

7.7   References

1.   Home PNA. "Simple, high-speed Ethernet Technology for the Home." White paper (June 1998).

2.   Home PNA. "News and Events." (March 2000).

3.   Bullock, Scott R. *Transceiver System Design for Digital Communications*. Atlanta: Noble Publishing, 1995.

8

Radio Frequency Communications

Radio frequency communications means sending and receiving signals using electromagnetic waves through the air. The RF communication systems are the only systems that are truly portable and wireless. Many applications require portability and wireless connections. RF communications can be used to bring signals, such as voice, data, audio, and video from the source that generates these signals to the home and distribute these types of signals through the home without running wires.

Wireless RF communications is becoming an important portable gateway to the Internet and other information sources, which allows the user to access information without being confined to specified locations. RF can be used in conjunction with other systems such as home PNA and PLC to enhance the coverage and add portability. RF Communications can also be used for mobile applications.

One of the drawbacks of wireless RF communications systems is the cost for high frequency parts. In the future, technology, including *application specific integrated circuits* (ASICs), the increased demand for high frequency parts, and competition between companies and manufacturers will allow for the cost of manufacturing RF systems to decrease.

RF communications is becoming a viable solution for bringing the information to the home and for distribution inside the home. Several of the standards to distribute RF in the home and to develop RF wireless networks are included in this chapter.

8.1   Basic RF Systems

RF systems require a transmitter for sending and transmitting information signals to a receiver. The transmitter uses an antenna in order to radiate enough energy to transmit these signals and overcome the losses, such as free space loss, atmospheric losses, and multipath through the air. The receiver is designed for sensitivity to detect low-level signals that come from far distances and for low noise to extend the dynamic range of the system. Also, power control, antenna diversity, and adaptive processes are often used to help overcome multipath, attenuation, jammers, and provide access for multiple systems.

Link Budget

A link budget is used to analyze the RF system and determine the amount of power, type of antennas, receiver sensitivity and noise figure, and the required signal to noise ratio S/N, or bit energy to noise power spectral density ratio Eb/No, and the link margin in order to send reliable information from the transmitter to the receiver. A typical link budget is shown in Table 8-1.

The receiver also has an antenna for receiving the transmitted signal. The transmitter is responsible for accepting the information, formatting, and modulating these signals on a carrier frequency. If the information is sent directly, the antenna would be very large and impractical. Therefore, the carrier frequency is the high frequency that carries the information through the air. The information must be modulated on the carrier (see Chapter 3). Most systems today use digital modulation techniques for optimal performance, security, versatility and cost. The receiver also uses a *low-noise amplifier* (LNA) in the front end to help establish the noise level of the receiver and provide the optimal *signal-to-noise* (SNR) for the detector. The receiver is responsible for receiving, demodulation, detecting and reliably recovering the information that was sent by the transmitter.

Antenna Size

Most of the RF systems that are used for bringing and distributing signals for the home use high frequencies so that the antenna lengths are short. Antennas are usually designed for a length that is equal to half the wavelength of the frequency of operation. The RF solutions generally operate in the open ISM bands that include the 900 MHz, 2.4 GHz, and 5.6 GHz bands, for distribution in the home, and even higher frequencies are being used for bringing the information to the home. To calculate the length of a typical antenna, the speed of light is divided by the frequency of operation to obtain the wavelength. A simplified formula takes the speed of light in millions of meters per second and divides it by the frequency in MHz (millions of cycles per second). The result is divided by two to solve for the length of the antenna that is equal to half the wavelength in meters.

$$\lambda = \frac{300}{f(MHz)} \quad Antenna\,Length = \frac{\lambda}{2} \qquad\qquad 8.1$$

Therefore, the half wavelength antennas for 900 MHz to 5.6 GHz converting to inches range from 1 to 6.6 inches. Smaller antennas, called loaded antennas, can be designed. Inductance is used to increase the electrical length of the antennas. Also, parabolic dishes are used for the higher frequencies to provide gain in the direction of the antennas.

Table 8-1 *Link budget for RF systems.*

Link Budget Analysis:							
	Range (m)	Freq.(GHz)	Power (mW)	Conversions:			
Enter Constants.......	100.0	2.4	100.0	Range:	Enter Range		Meters
					328.0	ft	100.0
Enter Inputs............	Inputs	Power Levels			0.06	miles	100.0
Transmitter	Gain/Loss (dB)	Sig.(dBm)	Noise(dBm)				
Tran.Pwr(dBm)=		20.0			Enter Power		Pwr(mW)
Trans. line loss =	-0.5	19.5		Power:	20.0	dBm	100.0
T/R switch	-1.0	18.5					
Trans Ant Gain =	2.1	20.6			Gain (dB)	Ant.	
ERP		20.6		Tx Ant Gain:	2.1	Dipole	
Channel				Rx Ant. Gain:	2.1	Dipole	
Free Space Loss =	-80.0	-59.4					
Atm loss(etc) =	-0.5	-59.9		Calculated Constants:			
Multipath Loss=	-20.0	-79.9		$\lambda =$	0.1	meters	
Receiver				$\lambda =$	4.9	inches	
RF BW(MHz)**	83.0		-94.8	Free Space Loss	-140.0	dB	
Rx Ant Gain =	2.1	-77.8		Boltzman's =	1.38E-23		
T/R switch	-1.0	-78.8					
Rec. Line loss =	-0.5	-79.3		Modulation	Eb/No(dB)	Eb/No	Pe
LNA Noise Fig. =	-1.0		-95.8	Coh.PSK	10.5	11.3	1.00E-06
LNA Gain =	20.0	-59.3	-75.8	Coh.FSK	13.5	22.6	1.00E-06
LNA levels =		-59.3	-75.8	Non.(DBPSK)	11.2	13.1	1.00E-06
Receiver Gain =	45.0	-14.3	-30.8	Non.FSK	14.2	26.2	1.00E-06
Imp. Loss	-2.0	-16.3		Coh.QPSK	10.5	11.3	2.00E-06
Gp for SS	0.0	-16.3		Coh. MSK	10.5	11.3	2.00E-06
Detect BW(MHz) **	20.0		-37.0	Non. DGMSK	14.2	26.2	2.00E-06
Det. Levels =		-16.3	-37.0				
S/N =		20.6					
Req. Eb/No							
Req. Eb/No		14.2	Pe=2.00E-06				
Coding Gain =	2.0	12.2					
Eb/No Margin =		8.4					

** Enter the actual bandwidths in MHz in the column even though they
are not gains and losses. The noise power is adjusted accordingly.
Note: The noise figure of the receiver is assumed to be the noise figure of the LNA.
This is not always the case, but it is generally a good approximation for a good receiver design.

RF Bandwidth

RF provides the required bandwidth to extend distribution of all types of narrowband and broadband signals and applications to multiple users at the same time. Some of the applications include high-speed Internet distribution and sharing, VoIP, HDTV, CD quality music, and other broadband signals.

Modulation Schemes

Several different types of modulation schemes are possible. A spectrally efficient direct sequence modulation scheme, for example, utilizes *Gaussian minimum shift keying* (GMSK). MSK modulation is a unique type of modulation that can be generated by either using a minimally spaced FSK or sinusoidally smoothed OQPSK (see Chapter 3). A Gaussian filter is added to reduce sidelobes to make the resultant waveform spectrally efficient.

Another method of modulation that increases data capacity for a specified bandwidth uses QAM systems. QAM utilizes both phase states and amplitude states to provide more data bits per symbol sent. Since the bandwidth is determined by the symbol rate, the more data bits per symbol the more information is sent for a given bandwidth.

Other modulation schemes include FH and *direct sequence* (BPSK, QPSK, 8PSK). For simple and less expensive systems, and to minimize the effects on variations over time, differential techniques are used. To increase the data throughput of a single channel, several parallel schemes are used to send data in parallel channels, such as OFDM.

Multiple user schemes, such as TDMA, FDMA, and CDMA prevent interference between users. Forward error correction FEC schemes and adaptive processes are also frequently employed to minimize the errors and reliability of the different systems.

RF Implementations to Reduce Inherent Problems

RF provides a wide range of coverage throughout the home and allows for portable wireless devices to access the home network. RF technologies have problems with multipath, blockage, and band saturation from other users, causing nulls in the amplitude spectrum and distorted signals. Power control, antenna diversity, and spread spectrum systems can help to mitigate these problems.

Power Control
Power control is very useful in CDMA type schemes where the process gain is limited due to short codes and limited bandwidths. It is used for many

types of systems to solve the near-far problem inherent in most systems. Generally, the closer the source is to the device, the stronger the signal is. If a device is very close to the source, and another device is far away, the near user will saturate or interfere with the user that is far away (see Figure 8-1).

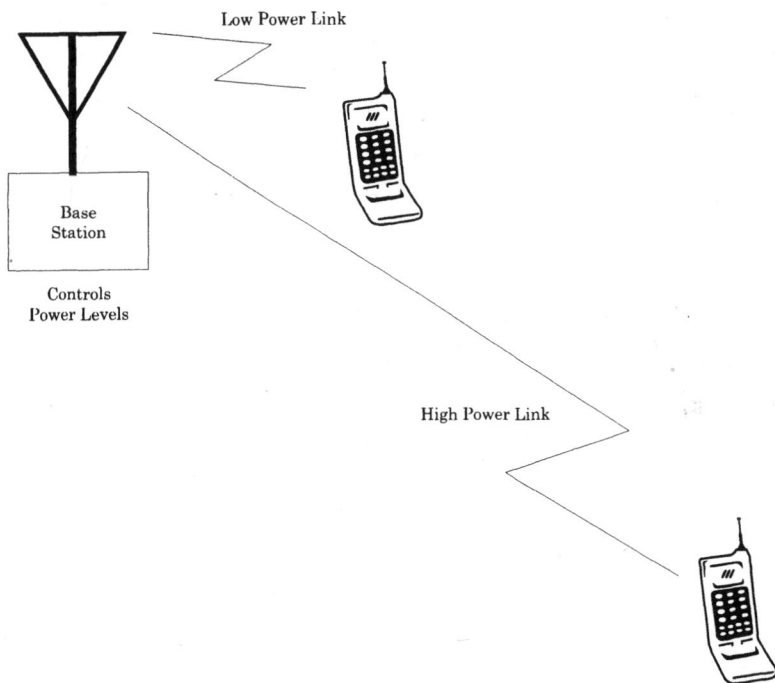

Figure 8-1 *Power control to mitigate the near-far problems.*

The power control operates by a request from the source to the user to transmit only the power required for a reliable link. Therefore, the near user would turn down the power output to the required level; the far user would turn up its power so that ideally, the power into the source would be approximately equal. Therefore, if a system has a limited process gain using CDMA or other multiple access schemes, with the power being equal due to power control, the jamming margin required to operate both systems can be optimized.

Antenna Diversity
RF wireless systems have problems with multipath and blockage or attenuation caused by obstructions, which can cause nulls in the band and make the communication link inoperable. Since the wavelength for most of these systems is very small, the multipath nulls are generally steep and nar-

row. A short distance of an inch or two can make the difference in trying to operate in a multipath null or in a desired part of the band. Since these systems are high frequency and short wavelengths, an additional antenna separated by a short distance can improve the reliability of the link tremendously. Either antenna can be used, depending on the quality of the signal or level of the signal into the antennas (see Figure 8-2).

Difference in separation reduces multipathproblems

Base
Station

Figure 8-2 Antenna diversity improves reliability.

The level in each of the antennas is monitored, and the best antenna is selected for operation of the link. This dynamic system is changing due to movement of the system and objects around the system. Other combining techniques can be implemented to reduce hardware and cost. Antenna diversity improves performance and reliability for stationary and mobile systems.

Spread Spectrum Technology

Spread spectrum systems are systems that use more bandwidth than is required to send the data to achieve process gain or jamming margin over other users or interferers. For PSK systems, the process gain is calculated by chips-per-bit. The chips are generated by a pseudo-random (pseudo-noise PN) code. The bits are the data or information that is sent. How fast the PN code is sent will determine how many PN chips are sent for every bit of information (see Figure 8-3).

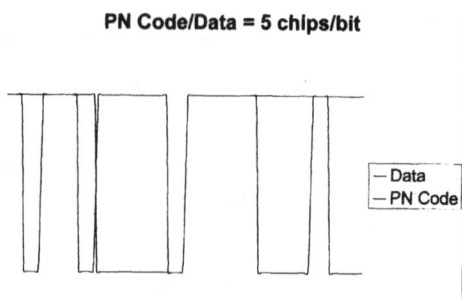

PN Code/Data = 5 chips/bit

—Data
—PN Code

Figure 8-3 Spread spectrum systems reduce the effects of other signals.

The process gain is calculated by the following:

$$\text{Process gain} = P_g = 10\log(\#\text{chips}/\#\text{bits}) \qquad 8.1$$

The actual performance of the system against jammers is called *jamming margin*. The jamming margin of a spread spectrum system is calculated by taking the process gain and subtracting both spreading losses and implementation losses as follows:

$$\text{Jamming margin} = M_j = P_g - L_s - L_i \qquad 8.2$$

where

L_s = spreading losses,
L_i = implementation losses.

For FH systems, the process gain is equal to

$$P_g = 10\log(\text{number of independent, hopped frequency channels}) \quad 8.3$$

For parallel multiple channel systems like orthogonal frequency division multiplexing OFDM, the process gain is defined as

$$P_g = 10\log(\#\text{parallel channels}) \qquad 8.4$$

The process gain associated with parallel channels means that the same data is sent on multiple channels. If the parallel channels are used to send different data to generate an overall higher data speed, then the process gain does not apply. Most systems use parallel channels to increase the overall data throughput or payload, not for process gain.

All of the above spread spectrum systems use more bandwidth than is required to send the data. The process gain is the ratio of the bandwidths. It is converted to a log power number by taking the log of the ratio and multiplying by 10. During the despread process, the noise is reduced with the despread code. Any jamming signals will be spread out and their amplitude will be lowered (see Figure 8-4).

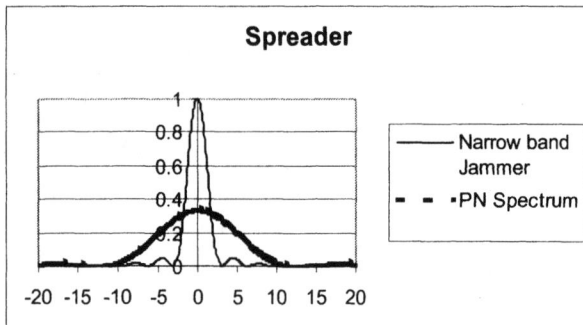

a.

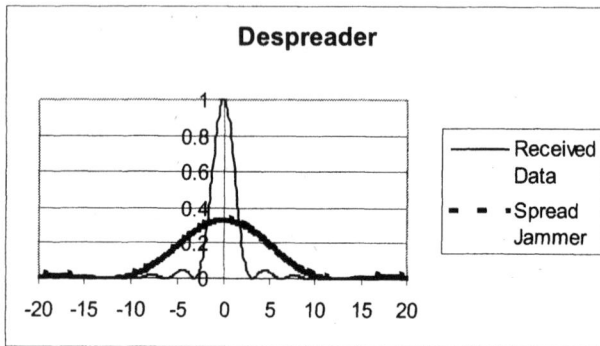

b.

Figure 8-4 *a. Spread signal and narrow band jammer, b. despread signal and spread jammer.*

8.2 Networking Standards for RF Systems

Networking standards are being established and verified in the areas of RF. Some of them so far are the IEEE 802.11 standard, Bluetooth, HomeRF, and HIPERLAN. Other RF standards are are discussed in Chapter 9.

IEEE 802.11

IEEE 802.11 standard specifies both the physical layer and the *medium access control* (MAC) layer and has been adopted as a standard for RF networking in the home. The physical layer can use both RF and *infrared* (IR). The RF uses spread spectrum techniques, including either FH or direct sequence and operates in the 2.4 – 2.4835 GHz ISM band. The 802.11 standard specifies data rates of either 1 Mbps or 2 Mbps.

The MAC layer uses carrier sense multiple access with collision avoidance (CSMA/CA) protocol. This allows multiple uses in the network and reduces the chance of transmitting at the same time. Each node checks to see if the channel is being used, and if not, it transmits it data. If the channel is busy, then the node randomly selects a specified amount of time that it waits before trying to transmit again. Since there are multiple time selections, the probability that two nodes trying to enter the network will select the same random time slot is very low. Thus, the chance of collisions or collision avoidance is reduced.

To ensure that a connection is made between two nodes, the transmitting node sends out a *ready-to-send* (RTS) packet and the receiving node upon successful reception of the RTS packet, sends a response of *clear-to-send* (CTS), which lets the transmitting node know that it is communicating with that particular node and not communicating with another node. Then the transmitting node sends the data, and the receiving node acknowledges reception by sending back an *acknowledge* (ACK) to the transmitter if it received the correct data. To verify that the received data is correct, the system uses a *cyclic redundancy check* (CRC).

Error Detection using Cyclic Redundancy Checking (CRC)

One of the best methods for error detection is the *cyclic redundancy check* (CRC). This method uses the division of the data polynomial $D(x)$ by the generator polynomial $G(x)$ with the remainder being generated by using an 'XOR' function instead of subtraction. The final remainder is truncated to the size of the CRC and attached to the message for error detection. The data polynomial is multiplied by the number of bits in the CRC code in order to provide enough place holders for the CRC. In other words, the remainder has to be large enough to contain the size of the CRC.

The polynomial is simply a mathematical expression to calculate the value of a binary bit stream. For example, a bit stream of 10101 would have a polynomial of $X_4 + X_2 + X_0$. The numeric value is calculated by substituting $X = 2$, $24 + 22 + 20 = 21$. Every '1' in the binary number corresponds to a value in the polynomial, with the power equal to the value it has in the digital number.

The receiver performs the same operation using the complete message including the attached error detection bits and dividing it by the generator polynomial. If the remainder is zero, no errors were detected. For example, suppose we have a data polynomial and a generator polynomial as follows:

$$D(x) = X_6 + X_4 + X_3 + X_0 \quad 1011001 \qquad 8.5$$

$$G(x) = X_4 + X_3 + X_2 + X_0 \quad 11101 \qquad 8.6$$

To provide the required place holders for the CRC, $D(x)$ is multiplied by X_4.

$$X_4(X_6 + X_4 + X_3 + X_0) = X_{10} + X_8 + X_7 + X_4 \quad 10110010000 \qquad 8.7$$

Note: This process is equivalent to adding the number of zeros as place holders for the size of the generator polynomial (four zeros).

The division using XOR instead of subtraction is

```
             1111110
11101|10110010000
      11101
      010110
      11101
      010111
      11101
      010100
      11101
      010010
      11101
      011110
      11101
      000110
```

Therefore, the CRC would be 0110 and the digital signal that would be transmitted would be the data plus the CRC, i.e., equal to 10110010110. The receiver would receive all the data including the CRC and perform the same function.

```
             1111110
11101|10110010110
      11101
      010110
      11101
      010111
      11101
      010100
      11101
      010011
      11101
      011101
      11101
      000000
```

Since the remainder is zero, there are no errors detected by the CRC. If there is a remainder, then there are errors in the received signal.

OFDM and CCK

OFDM algorithms have also been generated for the IEEE's 802.11b *wireless LAN* (WLANs) working group. OFDM allows for faster data rates using parallel techniques and orthogonal techniques to provide minimum adjacent channel interference. OFDM also allows the parallel frequency channels to overlap and still minimize adjacent channel interference to provide a spectrally efficient system.

The 802.11 working group developed a new modulation approach called *complementary code keying* (CCK) for the physical layer for data rates of 11 Mbps at 2.4 GHz carrier frequency. This new approach provides an easy path for interoperability with the existing 1 and 2 Mbps systems.

CCK is a variation of *m-ary orthogonal keying* (MOK) modulation, which uses an I/Q modulation. This modulation uses a *direct sequence spread spectrum* (DSSS) channelization scheme with three noninterfering channels in the 2.4 – 2.483 GHz ISM band. *M-ary bi-orthogonal keying* (MBOK) uses BPSK and 8-chip code words that are orthogonal. Since there are two phase states and 8 chips, the number of possibilities is $2^8 = 256$. CCK uses QPSK, which has four phase states and an 8-chip code word. Therefore, the number of possibilities for CCK is $4^8 = 65,536$. The codes are chosen so that they are nearly orthogonal to prevent interference between the code words.

The IEEE 802 working group set several standards relating to RF communication systems and are specified with an IEEE 802.xx. Many of the new RF systems pattern their designs based on the IEEE 802.11 standard.

Bluetooth

Bluetooth, IEEE 802.15, operates at 2.4 GHz in the ISM band (2.4 – 2.4835 GHz). The band allows 79 different channels with a channel spacing of 1 MHz. The band starts with 2.402 GHz that allows for a guard band of 2 MHz on the lower band edge and ends with 2.480 GHz, which allows a guard band of 3.5 MHz on the upper band edge (see Figure 8-5).

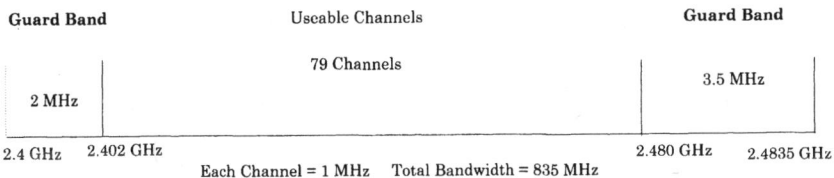

Figure 8-5 *Band allocation of Bluetooth with guard bands.*

Note that these band allocations and guard bands are the USA standards; other countries may operate at slightly different frequencies and with different guard bands.

There are three types of classes dealing mainly with power output and range: Class 1 – +20 dBm, Class 2 – +4 dBm, Class 3 – 0 dBm. The higher the power output, the greater the range of the system. However, both Class 1 and 2 require power control to allow only the required power to be used for operation. Class 1 operation is required to reduce the power less than 4 dBm maximum when the higher power output is not required. The power control uses step sizes from 2 dB minimum to 8 dB maximum. The standard method of power control uses a *received signal strength indication* (RSSI). The measurement from the RSSI in the receiver is sent back to the transmitter to adjust the transmitted power output.

The modulation scheme specified in the Bluetooth standard uses *differential Gaussian frequency shift keying* (DGFSK) with a BT = .5. The modulation index is between .28 and .35. A positive "1" is a positive frequency deviation, and a negative "1" is a negative frequency deviation with the frequency deviation greater than 115 kHz with a symbol rate of 1 Ms/s. Full duplex operation is accomplished using packets and TDM, which is often referred to as *time division duplexing* (TDD).

FH is used for every packet with a frequency hop rate of 1600 hops/sec providing one hop for every packet. The packets can cover one to five time slots, depending on packet size. Each slot time is 625 ms long, which corresponds to the hop rate. However, since the frequency hop rate is dependent on the packet size, if the packet is longer than the slot size, then the hop rate will decrease according to packet size. For example, if a packet covers five time slots, the hop rate would be one fifth of the 1600 hops/sec or 320 hops/sec, which equals 3.125 ms dwell time that equals five times 625 ms.

The standard packet is made up of three basic sections: access code, header, and the payload (see Figure 8-6).

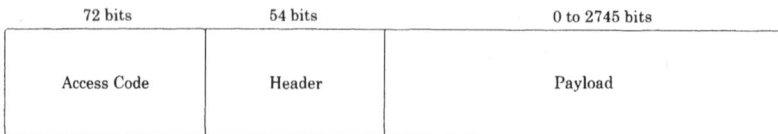

72 bits	54 bits	0 to 2745 bits
Access Code	Header	Payload

Figure 8-6 *Basic structure of a standard packet defined in the Bluetooth specification.*

The access code is 72 bits long and is used for synchronization, DC offset compensation, and identification. The detection method used is called a *sliding correlator*. Sliding correlators are used extensively in digital modulation schemes and spread spectrum technology. A code is generated by the transmitter and sent to the receiver along with the data. The receiver is slid in time against the received code. As the receiver code is slid past the transmitter code, an autocorrelation peak is generated if both codes line up within one chip time (see Figure 8-7).

Sliding Correlator   Auto-correlation Value

Transmit

Receive

Delay > 1 chip, early or late

Transmit

Receive

Delay = ½ chip late

Transmit

Receive

Delay = ½ chip early

Transmit

Receive

Delay = lined up

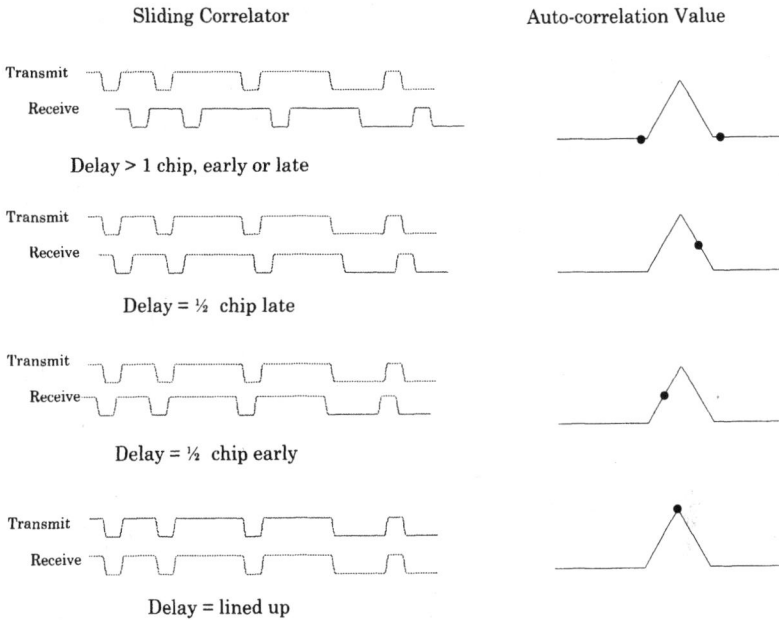

Figure 8-7 *Sliding correlator used for synchronization and code alignment.*

If the codes are not within one chip time, then the autocorrelation is minimal. If the codes are within half chip time, then the autocorrelation value is equal to half the maximum correlation. If the codes are perfectly lined up, then the maximum auto-correlation value is achieved. Many locking systems use the auto-correlation to synchronize the codes. They use a ½ early, and ½ chip late, and compare the auto-correlation values. If they are equal, then the system is synchronized. If the half chip early auto-correlation value is larger, then the oscillator frequency is increased until they are equal. If the half chip late autocorrelation value is larger, then the oscillator frequency is decreased until they are equal. This provides a feedback system to keep the receiver and transmitter codes locked. This method of keeping the codes in lock is called an *early-late gate.*

The three types of access codes are *channel access code, device access code,* and *inquiry access code.* The channel access code identifies the piconet. The device access code is used for applications such as paging. The inquiry access code is used for *general inquiry access code* (GIAC) and checks to see what Bluetooth devices are within range. The *dedicated inquiry access code* (DIAC) is used for designated Bluetooth units in a given range.

The header is 54 bits long and determines what type of data is being sent, for example, "link control." The payload is the actual data or information that is sent. Depending upon the amount of data, this can range from 0 to 2,745 bits in length.

Three error correction schemes are used with Bluetooth: one-third rate *forward error correction* (FEC), two-third rate FEC, and *automatic repeat request* (ARQ). Error detection is accomplished using CRCs. The standard also specifies a scrambler or a data whitener to prevent long constant data streams (all "1"s for a period of time or all "0"s for a length of time), which causes problems in carrier and tracking loops.

The capacity of the Bluetooth standard can support an asynchronous data channel, *asynchronous connectionless link* (ACL), three point-to-point simultaneous synchronous voice channels, *synchronous connection-oriented* (SCO) links, or an asynchronous data channel and a synchronous voice channel. The synchronous voice channel is 64 kbps for quality voice in both directions. The asymmetrical speeds for Bluetooth are up to 723.2 kbps and 57.6 kbps on the return channel and 433.9 kbps if both links are symmetric.

Bluetooth forms networks that are either *point-to-point* (PTP) or *point-to-multipoint* (PMP). The PMT is called a piconet and consists of a master unit and up to seven slave units. The slave units can be in multiple piconets, which is called a scatternet using TDM; however, the piconets cannot be time or frequency synchronized.

The dynamic range of the Bluetooth system is required to have a minimum dynamic range of 90 dB. The receiver for Bluetooth is required to have a minimum sensitivity of –70 dBm and needs to operate with signals up to +20 dBm.

The initial Bluetooth specification was noted for its short range and used mainly for device control applications. With the increase in power output, up to +20 dBm, the range has increased to where it can be used for multiple applications.

Networking in the home will include multiple technologies, each with their particular advantage. Bluetooth, because of the versatility and mobility of the connection, will have a position in the future of broadband and home networking systems, along with other RF systems, home PNA, and power line.

HomeRF Working Group

A wireless specification called the *shared wireless access protocol* (SWAP) has been developed by the *homeRF working group* (HRFWG) and supported by several manufacturers. This wireless standard operates in the 2.4 GHz ISM band. Its specifications are similar to the IEEE 802.11 standard and DECT. The transmit power is up to +24 dBm and uses FSK modulation for a data rate of 0.8 Mbps. Another modulation type uses a 4-FSK (four different frequencies possibilities that provide twice the data rate as standard FSK) to increase the data rate to 1.6 Mbps. This mode is also available under this standard. FH, with a hop time of 300 us is also used for spreading the signal, FHSS. HomeRF, like 802.11, can support over 100 nodes and up to four voice conversations. The output power is up to +20 dBm, which is adequate for most home installations. The HomeRF group is also looking at ways to develop standards to interconnect different types of wired technology, such as Home PNA, Ethernet, and power line. All networking applications are looking at voice, data, audio and video. Applications include printers, scanners, CD drives, DVD drives, multiple PCs, camcorders, Internet sharing, games, and many other applications. With regards to video transmissions, the IEEE 1394 is the expected standard.

HomeRF supports both isochronous devices that are slaves to the main computer and asynchronous network technology, which is a wireless Ethernet. The HomeRF group targets three applications: PC-enhanced cordless telephone (with many feature attached to tying telephony with the PC), mobile viewer appliance (which could use a device similar to a notebook type of computer to view applications and interface to the network), and resource sharing (such as using the same printer with multiple PCs, Internet sharing, games, and backup storage). These applications would be similar to the present wired networks, but could be extended to many types of peripheral devices and provide mobility and portability. The interactive voice applications use TDMA for multiple users operating the same technology and bandwidth. The asynchronous networking uses *carrier sense multiple access* (CSMA) with the information being transmitted in packets the multiple users detect to see if the channel is clear to transmit their information.

HIPERLAN

HIPERLAN was developed within the European Telecommunications Standards Institute (ETSI). ETSI set the GSM standards for digital cellular telephony. HIPERLAN's mission is to develop high-speed wireless data solutions for multimedia communications. Aligned with the IEEE 802 standards, the HIPERLAN Type 1 system resembles a wireless Ethernet running at speeds up to 23.5 Mbps, with a carrier frequency of 5 GHz.

Several standards are being developed for HIPERLAN. HIPERLAN 2 uses a centrally controlled system, with *asynchronous transfer mode* (ATM) technology running at 20 Mbps and with a carrier of 5 GHz. HIPERLAN Type 3 or HIPERAccess deals with the wireless local loop with the same data rate of 20 Mbps and the same carrier frequency of 5 GHz. Also, a high-speed point-to-point link is considered that would have very high data rates up to 155 Mbps and carrier frequency of 17 GHz under the standard HIPERLAN Type 4 or HIPERLink.

The HIPERLAN committee in ETSI identified the band of 5.15 GHz, with an option to increase the upper limit to 5.3 GHz, and a bandwidth of 100 MHz to cover the high speeds and the number of users expected.

HIPERLAN's modulation scheme is the proven GSM technology with GMSK modulation to achieve the high data rates for less cost that other type of technologies. HIPERLAN help established RF standards that are published by ETSI, including ETS 300, ETS 652 and conformance specification ETS 300, ETS 864.

8.3   Summary

RF communications is the only true wireless technology to connect high-speed communications to and throughout the home and to provide a network that is both portable and mobile.

Several standards are in existence or are being created to accommodate the tremendous technology growth curve to provide higher data rates and more universal home networks. There is an accelerating need to bring and distribute high-speed information in the home including voice, data, audio and video for applications such as providing an Internet gateway, feature rich telephony, CD quality music, shopping, games, file sharing and print sharing, memory storage, appliance and environmental controls, security systems, movies and television including HDTV, and many more future devices. Four standards that are well on their way in establishing true wireless connections include IEEE 802.11 standard, Bluetooth, HomeRF, and HIPERLAN. The RF communications systems will investigate other technologies such as Home PNA and PLC to try and bridge the gap to provide a complete networking system for the home.

The RF communication systems will provide wireless networking in the home that would interconnect many types of peripheral devices and provide mobility and portability.

8.4   References

1. Lough, Daniel L., T. Keith Blankenship and Kevin J. Krizman. "A Short Tutorial on Wireless LANs and IEEE 802.11." (March, 2000).

2. "Specification of the Bluetooth System." *Core* (December, 1999).

3. Andren, Carl and Mark Webster. "CCK Modulation Delivers 11 Mb/s For A High Rate 802.11 Extension." *Wireless System Design* (May 1999).

4. Taylor, Larry "HIPERLAN Type 1 Technology Overview." White Paper (June 1999), Rev. 0.9., TTP Communications Ltd.

5. Grewe, Tony and Rich Nesin. "Wireless Home Has Many Suitors." *Electronic Engineering Times* (November 29, 1999).

6. Negus, Kevin J. and Adrian P. Stephens, Jim Lansford. "HomeRF: Wireless Networking for the Connected Home." *IEEE Personal Communications* (February 2000).

7. Schneiderman, Ron. "Interoperability Tops Bluetooth Vendor Issues." *Supplement to Penton's Electronics Group* (winter issue).

8. Anderton, David O. and Steven W. Stanton. "Developing Measurement Solutions for Bluetooth." *Supplement to Penton's Electronics Group* (winter issue).

9. Bullock, Scott R. *Transceiver System Design for Digital Communications*. Atlanta: Noble Publishing, 1995.

10. Holmes, Jack K. *Coherent Spread Spectrum Systems*. New York: John Wiley & Sons, 1982.

11. Schwartz, Mischa. *Information, Transmission, Modulation and Noise*. New York: McGraw-Hill, 1980.

12. Haykin, Simon. *Communications Systems*. New York: John Wiley & Sons, 1983.

13. Bullock, Scott R. "Phase-Shift Keying Serves Direct-Sequence Applications." *Microwaves and RF* (December 1993).

9

The "Last Mile"

The "last mile" specifies the connection from the local distributor of the signal to the end user or home. Several companies are providing different methods to connect and provide services of voice, data, music, video, and all other communications. They include wired solutions PSTN, cable, and fiber optics, and wireless solutions, fixed wireless, power line, and satellite. Also, many of these technologies are combined to provide new methods of bringing the communications signals to the home, including hybrid fiber/copper, hybrid fiber/coax, and wireless/wired systems.

Many wireless systems in multiple bands compete for the "last mile." Methods considered include point-to-point, point-to-multipoint, and multipoint-to-multipoint for bringing broadband communications information and provide networking capabilities amongst the end users. The basic requirement for all of these methods is to bring high-speed information to the home and office environments.

9.1 Types of Transmission Mediums

There are five methods of bringing information to the home.

- telephone wire
- coaxial cable
- fiber optics
- wireless and RF
- satellite communications

There also techniques that use a combination of the above technologies for a complete system to provide optimal coverage of both stationary and mobile applications.

9.2 Telephone Wire

Telephone wired systems use copper wire to connect directly from the home to the CO utilizing unshielded twisted pair.

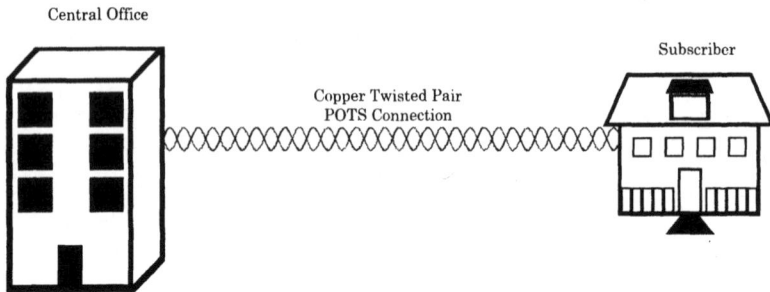

Figure 9-1 *Direct connection to the central office using twisted pair POTS telephone wire.*

The unshielded twisted pair consists of two wires twisted together to help noise immunity (PSTN or POTS). PSTN has been the only connection in existence for many years. Eventually, the PSTN connection will become obsolete as higher bandwidth, i.e., more reliable connections are developed and deployed.

9.3 Coaxial Cable

The coax shielded cable medium brings video into the home from the cable TV providers (see Figure 9-2). Many of the RBOCs have used the existing infrastructure to their advantage in bringing telephone connections via the coax cable. Coax cable has very good noise immunity and is more durable than the PSTN unshielded twisted pair medium.

Most of the existing wiring is referred to as CAT3. The newer CAT5 wiring is used with fiber optic cable and improves the crosstalk balance by 20 dB. The differential mode provides a crosstalk balance of approximately 55 dB, which provides a tremendous improvement to prevent interference from adjacent signals.

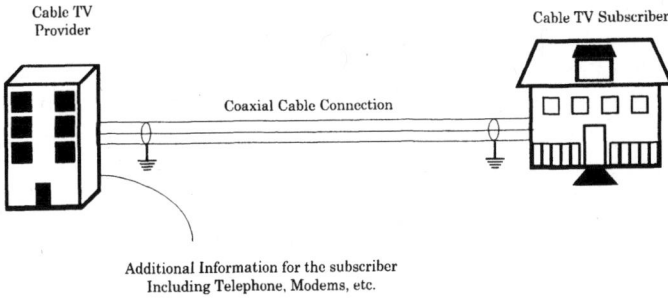

Figure 9-2 *Coaxial cable connection to the home using the cable TV providers.*

In North America, T1 lines have a data rate capability of up to 1.544 Mbps dual simplex on 2 pair of wires. A repeater is needed every 1000 feet to provide a boost of the signal so that longer ranges can be achieved. Outside North America, E1 lines are used, with data rates up to 2.048 Mbps with 2 and 3 pair full duplex systems.

9.4　Fiber Optics

Fiber optics provide a medium with very high bandwidth capabilities for higher speed data connections and more users-per-connection than standard telephone wired systems (see Figure 9-3). Fiber optics have very low loss compared to the hard-wired medium and are used mainly for long distances.

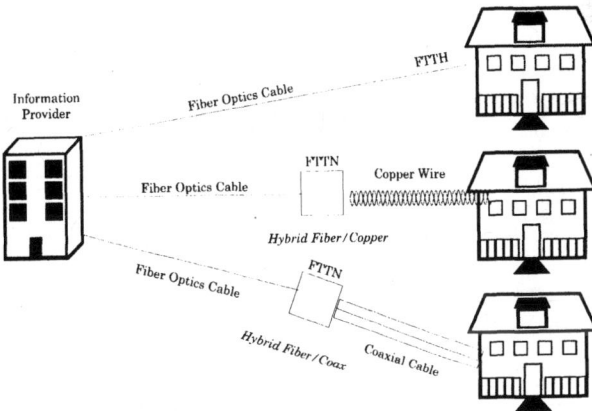

Figure 9-3 *Fiber optics connection to the home, hybrid fiber/copper, and hybrid fiber/coax.*

Fiber optics are becoming popular for hard-wired systems that require high bandwidths and high data rates. Fiber optics connected to the home are called *fiber to the home* (FTTH). If fiber optics cable is run to the neighborhood, which is quite common, and coaxial cable is run from the fiber optics point to the homes, it is known as *fiber to the neighborhood* (FTTN). When fiber optics cable is run to the exchange, like the central office, then it is called *fiber to the exchange* (FTTE). To achieve low loss, high reliability and high performance, fiber optics cable is by far the most superior for hard-wired systems. However, it is used with other hard-wired systems to form hybrid connections, hybrid fiber/copper, and hybrid fiber/coax.

9.5 Wireless and Radio Frequency

The wireless and radio wave connection uses electromagnetic waves to transport the telephone signals through the air (see Figure 9-4).

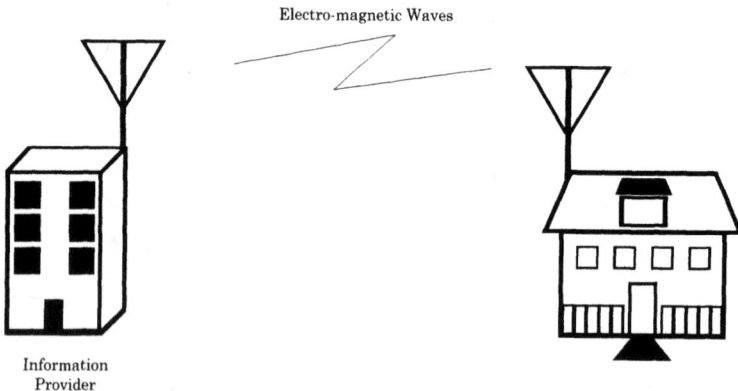

Figure 9-4 *Wireless and RF connections.*

Systems that utilize wireless radio wave transmission mediums are fixed wireless, microwave, satellite, cellular, PCS, pagers, and other hand-held products. There has been a tremendous effort over the last few years to establish an infrastructure to handle all of the wireless systems. Both analog and digital systems have been implemented. Analog cellular has been around for years, but as the PCS technology began to implement digital systems, the cellular companies produced digital cellular to compete in the market place. PCS uses spread spectrum technologies, such as TDMA, CDMA, FDMA, FH and GSM. GSM was adapted mainly to improve interoperability between different manufacturers. Combinations of the above technologies were implemented, along with making them cellular compatible.

9.6   Local Multipoint Distribution Service (LMDS)

Local multipoint distribution service (LMDS) is a "last mile" point to multipoint distribution service, which distributes communications signals with a relatively short RF range to multiple end users. In this multipoint system, the base station or hub transmits signals in a *point-to-multipoint* (PMP) method that resembles a broadcast mode. The return path from the subscriber to the base station or hub is accomplished by a *point-to-point* (PTP) link.

The LMDS system is an end-to-end microwave radio link that enables VoIP, local and long distance telephony services, high-speed Internet access and data transmissions, as well as video conferencing and TV.

LMDS uses digital wireless transmission systems operating in the 28 GHz range for systems that operate in the United States and 24 – 40 GHz for overseas systems depending on the area and country. LMDS can be employed in either an asymmetric or symmetric configuration. The asymmetric system has a data rate that changes depending on the direction of the transmission; for example, the downstream has a higher data rate than the upstream, as is the case with *asynchronous digital subscriber loop* (ADSL). The symmetric configuration has the same data rate regardless of the direction the data is sent.

Frequency allocations for LMDS are designated by the FCC. These allocations are assigned as blocks to the end user. Block A users are assigned 1150 MHz of bandwidth in different frequency bands, as shown below:

27.5 – 28.35 GHz	850 MHz
29.1 – 29.25 GHZ	150 MHz
31.075 – 31.225 GHz	150 MHz
Total Bandwidth =	1150 MHz

Block B users are left with the remainder of the bandwidth, which equals to 150 MHz total. This total bandwidth allocation of 1300 MHz for LMDS represents the largest bandwidth allocation by the FCC of the wireless data communications systems including PCS, Cellular, DBS, MMDS, digital audio radio services, wireless communications service, and interactive and video data, and over two times the bandwidth of AM/FM radio, VHF/UHF television, and cellular telephone combined. Most of the above systems are allocated at a fraction of the LMDS allocation; for example the ISM band for PCS at the 900 MHz band is only 26 MHz. The standard IF frequency for typical LMDS systems ranges from 950 to 2050 MHz.

Many LMDS systems utilize the concepts of bandwidth on demand and shared resources by integrating their systems with ATM. IP transport methodology is also used to send data and VoIP.

LMDS is used to span the "last mile" to the user's facilities. LMDS systems transmit data rates up to 500 Mbps each way. However, the distance is limited to about 2 to 4 miles because of the rain-fade at these frequencies of up to 30 dB of attenuation in the link analysis. Using the bandwidth allocated to the LMDS systems can provide up to 85 Mbps to the residential users and up to 155 Mbps to the commercial users. A typical LMDS can provide up to 155 Mbps downstream and a return link of 1.544 Mbps – standard T1 line. These high-speed data rates are compatible with the T1 speed requirements up to and including OC-3 connections. Using just the first part of the Block A bandwidth of 850 MHz and using QPSK modulation, this system can provide 100 simultaneous T1 lines. With the additional bandwidth and higher order modulation schemes, this capacity can be increased.

All communication systems that are involved with providing multiple users to access a communication system require a method of multiplexing these users. The basic multiplexing schemes used are code division multiplexing/multiple access (CDM/CDMA), time division multiplexing/multiple access (TDM/TDMA), and frequency division multiplexing/multiple access (FDM/FDMA). The difference in the multiplexing methods and the multiple access methods is that the multiplexing method assigns the code, time slot, or frequency to a user regardless of whether they are using the system or not. In contrast, the multiple access schemes assign the user a code, time slot, or frequency on demand and when they access the system. This assignment is only for the period of that time of use, then that code, time slot, or frequency is released and made available for the next user after the original user exits the system.

Most LMDS systems today use either FDM/FDMA and/or TDM/TDMA to provide multiple access for a given system, depending on needs and usage. FDMA works well for applications where the user is connected on a more continuous basis; TDMA works well when the user is more periodic and the demand is on an as-needed basis and less continuous. If two users require the link on a continuous basis, TDMA would not be suitable because the users would have to share the time allocation; whereas, with FDMA, the time is always allocated to each of the users, and the access is separated in frequency. However, if the two users are periodic in there usage, and do not require continuous access to the system, then TDMA might be the method to use. TDMA takes advantage of the time when the subscriber is not using the services and allocates this time to another subscriber to either increase the number of users on the system or increase the overall data rate of the connection. These factors and others are considered when deciding on which type of multiplexing is optimum. Frequently, both TDMA and FDMA are used in a system to provide better access, more users, and higher speeds for each user with minimal interference between them.

FDMA may be the best solution to large customers who use the Internet on a fairly continuous basis. Small end users require high usage for downstream applications while they are downloading large files, but generally the usage and speed requirements are lower during upstream use. Therefore, to provide an optimal system, a TDM/TDMA scheme is generally used.

Other methods of providing multiple users include orthogonal methods and spacial antenna separation. Some examples of orthogonal techniques include OFDM, orthogonal phase systems like QPSK, and orthogonal antenna polarizations, which, for example, use vertical and horizontal polarizations for different users. This method is currently being used in satellite communications and in other communication systems.

Spacial antenna separation, or space sectoring is another multiple access scheme that allows multiple subscribers to use the same communications system by having directional antennas pointed to different space segments of the intended users. The number of users and spacial separation depend on the beamwidth and the amount of isolation between beams. This allows multiple subscribers to use the same system without jamming each other. Some of these methods are used in various combinations of orthogonal techniques, multiple access schemes, and spacial antenna separation to provide the optimum solution and the maximum amount of end users per system.

LMDS use different types of digital modulation schemes, depending on complexity and bandwidth efficiency. Some of the more popular modulation schemes include BPSK, QPSK, QPSK, 8PSK, using both standard and differential on any of the modulation types, and various QAM systems including, basic QAM, 16-QAM, and 64-QAM. The higher the order of modulation scheme, the higher the data rates for a given bandwidth. However, these systems are generally more complex, require higher signal levels (or have a reduce range), and have a higher cost of implementation.

Modulation techniques provide a higher data rate for bandwidth efficiency, and multiple access techniques allow multiple users to access the system. The combination of these methods is used to maximize the capacity of a given system by providing the maximum users per base station or site. The higher capacity that can be achieved for a given base station provides more coverage or less base stations per given coverage. Therefore, the modulation and multiple access schemes are carefully designed for each system installed.

Since the LMDS has a relatively short RF range, the base stations or hubs are spaced a few kilometers apart and linked together to provide service up to several thousand end users. LMDS systems are affected by atmospheric conditions, mainly rain, which is the main cause for reduced coverage in these microwave systems. LMDS systems are line-of-site LOS, antennas are fixed at the sight, they are usually mounted at a high elevation (rooftops), and often the antennas are directional, so multipath and blockage is generally not a problem after the antennas are mounted and operational. However, the atmospheric affects are constantly changing and affect the range of the system. Antenna siting and mounting are important factors along with the type of modulation used to provide the range required for coverage and capacity.

An LMDS system consists of four elements: the *customer premises equipment* (CPE), the base station or hub which services multiple end users, the fiber-based infrastructure which is the wired connection to the CO, and the *network operations center* (NOC) which provides the networking and can operate with or without a CO.

The CPE has many different types of devices that are used by the end user. They include LANs, telephones, faxes, Internet, video, television, set top boxes STB, and other devices. The interfaces to these devices include; digital signal, DS-0, DS-1 structured and unstructured T1/E1, T3/E3, DS-3, OC-1, OC-3/STS-3 fiber optics, ATM and video communications, POTS, frame relay, Ethernet 10BaseT/100BaseT and others. A *network interface unit* (NIU) at the CPE is used to interface between the incoming LMDS signals and the devices that are being used at the CPE. There are scalable and non-scalable NIUs. The scalable NIU is used in large businesses and for commercial uses. It is a flexible system that can be configured for the application and is chassis based. Therefore, the same chassis can be used for many types of systems and applications.

The main elements of the NIU are the modem and the data processor that supports the different type of external connections or interfaces. The nonscalable unit is for small and medium sized users, for a fixed application and interfaces. This unit is designed specifically for certain types of interfaces, T1/E1, T3/E3, POTS, 10BaseT, video, frame relay, and ATM.

The *network-node equipment* (NNE) connects the wireline functions to the LMDS wireless link. This equipment contains the processing, multiplexing/demultiplexing, compression/decompression, modem, error detection and correction, and ATM.

The base station or hub provides the interface between fiber infrastructure and wireless infrastructure. The elements of a standard base station are antenna system and microwave equipment for both transmitting and receiving, downconverter/upconverters, modulators/demodulators, and interface to the fiber. The base station antennas are mounted on rooftops and other high places to prevent blockage from structures, since the LMDS frequencies are *line-of-sight* (LOS).

Local switching in the base station can permit communications between users operating in the same network without going through the wired infrastructure if a NOC is provided. The fiber-based infrastructure is the wired connections mainly to the CO, which provides the connections from the base stations or NOC to the local CO.

The NOC consists of the *network management system* (NMS) equipment consisting of optical network (SONET), optical carrier OC-12 and OC-3, DS-3 links, central office CO equipment, ATM and IP switching systems, and interconnections with PSTNs. Some implementation schemes do not use a NOC; instead the information from the base station is connected through fiber to ATM switches or CO equipment at the CO, so that all communications must go through the CO. By using a NOC, if users on the same network desire to communicate, they can do so directly through the NOC and bypass the CO.

The current applications for LMDS are used in locations where it becomes difficult and impractical to run copper wire. Since LMDS is a wireless solution, it will have a minimum impact to the community, and at the same time provide high-speed communications through its extended bandwidth.

For rural communities, LMDS offers a solution by providing a wireless link to these remote communities at a lower cost and less time to install. In remote countries LMDS can provide a solution by setting up a wireless infrastructure that is less costly and requires less time to install. LMDS can not only provide an immediate solution to the problem, but can extend the coverage that the wired lines provide. Presently, most of the applications are dealing with businesses due to the high cost of implementation for home use. However, over the next several years, this will become more viable for service to the home.

9.7   Microwave (or Multichannel) Multipoint Distribution Service (MMDS)

MMDS is a wireless service that operates from 2.2 to 2.4 GHz. It has a range of approximately 30 miles, which is line-of-sight at these frequencies. It was originally designed to provide cable TV to subscribers in remote areas or in locations where it is difficult to install cable. Other systems use bandwidths of 200 MHz in the band just above 2.5 GHz and also in the Ka-band at 24 GHz. The power output allowed is up to 30 watts, and OFDM is used to enhance the number of users and increase the speeds. This provides up to 10 Mbps during peak use and can provide speeds up to 37.5 Mbps to a single user.

MMDS has the capacity to support up to 33 analog channels and more than 100 digital channels of cable television. In 1998, the FCC passed rules to allow MMDS to provide data and Internet services to subscribers. MMDS is mainly a short-range, inexpensive solution.

9.8   Standards for LMDS and MMDS

The new IEEE 802.16 working group for broadband wireless access networks is working on both the MMDS and LMDS and other wireless "last mile" type technologies. Both OFDM and VOFDM technology has been proposed for use in the MMDS system solution. The lower frequency bands, 2.5 and 5.0 GHz are more appealing at the present time because of their low cost, easy availability and fast installment. LMDS may provide a long-term future solution to the "last mile" connection but it will take more time to get the infrastructure in place for this line-of-site technology. Other groups that are involved in setting standards for LMDS and MMDS are the International Telecommunications Union (ITU) and the European Telecommunications Standards Institute (ETSI).

9.9   Satellite Communications

Satellite communications provide video programming to the home using *set top boxes* (STB) and satellite antenna dishes. One of the first systems to emerge was the Iridium system, which offered voice and low-rate data services. The Iridium systems was inherently impractical and was never produced in massive quantities.

Broadband systems, with high-speed data connections, voice and video, operate in the C, X (mainly military), and Ku bands. Newer systems are being designed to operate in the Ka-band with new satellites offering Ka-band service. Satellite systems could replace some of the terrestrial systems, and enhance the systems by providing coverage in remote places.

Satellite communications offer the most complete infrastructure and coverage compared with any other technology. Satellite communications is established by converting voice, data, video to a digital format, upconverting to a higher frequency to transmit up to the satellite. The satellite then transmits down to a ground-based receiver at the provider of these signals. If the satellites act only as a space transponder, then they are commonly called a "bent pipe." These signals and the ground stations are downconverted, demodulated and detected to receive the information sent (see Figure 9-5). The path back operates in the same way. The signals are upconverted from the provider and sent to the satellite. The satellite then sends the information to the subscribers' receiver where it is downconverted, demodulated, and detected and the information is received.

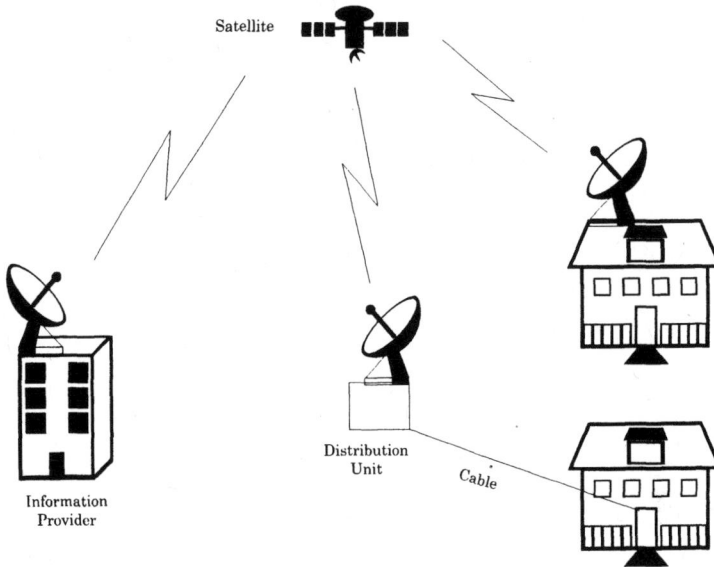

Figure 9-5 *Satellite communications link to remote areas.*

9.10 Summary

Many technologies can complete the final connection and distribution to the home or office, i.e., the "last mile." Ubiquitous infrastructure, manageable costs, high-speed information, multiple users and high traffic, broadband communications, and networking are the major design criteria for providing the "last mile." Approaches to the "last mile" include hardwired systems, wireless systems, including RF and microwave technologies, and satellite systems. In order to achieve all of these design criteria, multiple solutions may be combined for a complete solution.

Two major emphases for broadband and home networking is the method or methods of bringing the information to the end user, and the other is distribution of this information inside the home of office. Five methods were discussed that provide the information to the user. Three methods of distribution were discussed. It may be a combination of many of the technologies proposed, and maybe even some new proposals. New standards are being reviewed as new technologies arise and as the data rates increase.

9.11 References

1. Wirbel, Loring. "LMDS, MMDS Race for Low-Cost Implementation." *Electronic Engineering Times* (November, 1999).

2. "Satellite Communications Handbook, Fixed Satellite Service." *International Telecommunications Union, International Radio Consultative Committee*, Geneva (1988).

3. Bullock, Scott R. *Transceiver System Design for Digital Communications.* Atlanta: Noble Publishing, 1995.

4. Inglis, Andrew F. *Electronic Communications Handbook.* New York: McGraw-Hill, 1988.

5. Bullock, Scott R. "Use Geometry to Analyze Multipath Signals." *Microwaves & RF* (July 1993).

6. Bullock, Scott R. "Phase-Shift Keying Serves Direct-Sequence Applications." *Microwaves and RF* (December 1993).

7. Fuertes, Andy. "Opportunities and Challenges Facing LMDS." *America's Network* (June, 1998).

8. Mason, Charles. "LMDS: Fixed Wireless Wave of the Future?" *America's Network* (June, 1998).

9. "Local Multipoint Distribution System (LMDS) Tutorial." *The International Engineering Consortium Web ProForum* (August 3, 2000).

11. Tipparaje, Vinod. "Local Multipoint Distribution Service (LMDS)." Ohio State University (December 2000).

10

Satellite Communications

Satellite communications allows the most remote places to receive Internet, telephones, faxes, video and telecommunications via satellite connections. The infrastructure, bandwidth, and availability of satellite communications, and the possibility of combining satellite communications with other types of communications systems makes this method an ideal candidate for providing ubiquitous communications to everyone worldwide.

10.1   Communication Satellites

Communication satellites, which consist of a space platform and the payload, are positioned in a geostationary orbit. The payload is the equipment and various other devices that are mounted on the space platform. The same satellite may contain multiple payloads for multiple applications, including different data links operating with their own frequencies and antennas.

Each satellite is equipped with attitude control to ensure that the antennas are all pointed towards the earth and that the antenna beam is focused on the intended area on the earth (see Figure 10-1).

Any disturbances that may occur will change both the attitude of the geostationary satellite and the orbit of the satellite by a slight amount. The attitude stabilizers that are on board the satellite compensate for the changes in attitude, which is the positioning of the satellite for the antennas. The orbit of the satellite is changed due to movements in the north and south plane in a figure eight pattern over a twenty-four hour period.

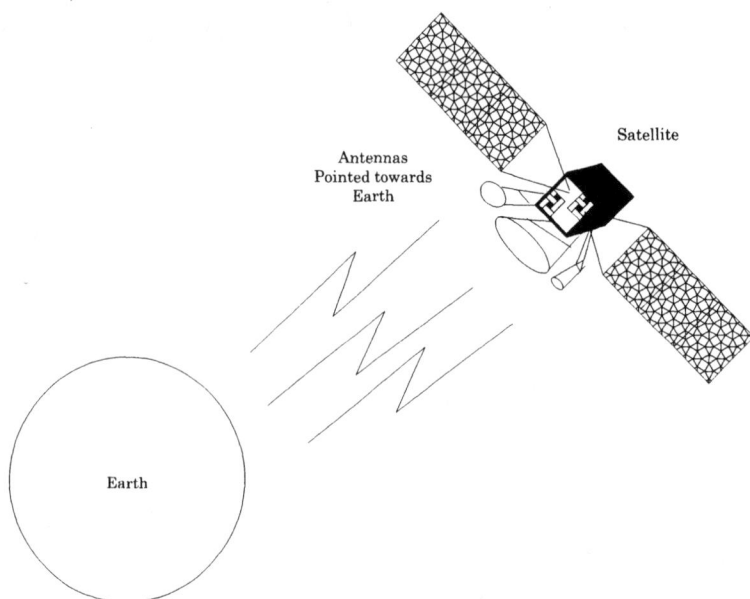

Figure 10-1 *Attitude control to ensure the antenna's pointing direction.*

10.2 General Satellite Operation

End users are connected to an earth station at their location, whether it is a personal system or a network serving multiple users. This earth station provides two-way communications up to the satellite. The space station relays the two-way information to another earth station that provides the services, whether it is Internet services, telephone, fax, video or music (see Figure 10-2).

Operational Frequencies

Four different bands of operation are used for satellite communications. The lowest band is called the *L-band*. It operates with an uplink at 1.6 GHz and a downlink at 1.5 GHz using a narrow bandwidth. The next band is the *C-band* and operates around 6 GHz for the uplink to the satellite and 4 GHz for the downlink from the satellite to the ground station. The next band, which is generally used by the military, operates in the X-band and operates around 8 GHz for the uplink and 7 GHz for the downlink. The next band, which has become popular for telecommunications, is the Ku-band and operates around 14 GHz for the uplink, and 11 to 12 GHz for the downlink. The highest band of operation, the Ka-band, is becoming popular for broadband communications and other applications. This band operates at 30 GHz for the uplink and 20 GHz for the downlink. The Ka-band provides a much higher bandwidth for high-speed data and allows for more simultaneous end users. A summary of the different frequency bands, including the bandwidths are shown in Table 10-1.

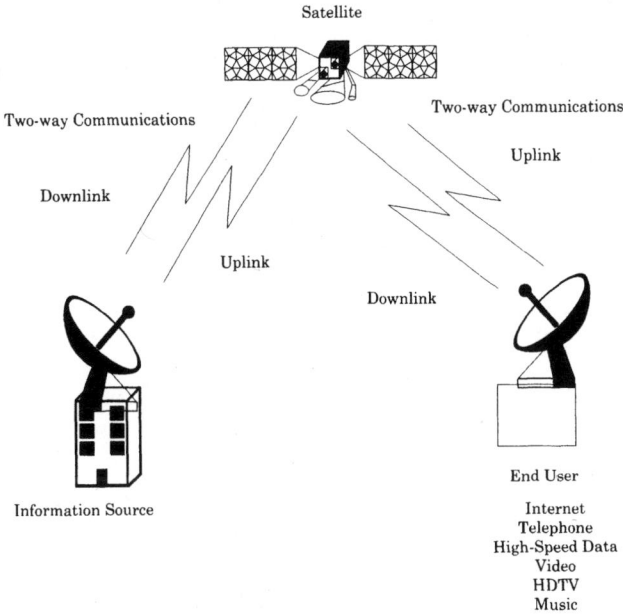

Figure 10-2 *Satellite operational communications link.*

Table 10-1 shows approximate frequencies that represent bands that are in use. Band expansion is continually changing so these numbers may not be exact and may vary slightly. Along with frequency changes, bandwidths also change for different systems and uses. The most widely used bands are the 6/4 GHz C-band and the 14/11 – 12 GHz Ku-band. However, the 30/20 GHz band has become very popular in recent years for commercial communications and is growing rapidly due to the interest in broadband communications.

Table 10-1 *Satellite frequency bands.*

Frequency Band	Uplink	Downlink	Bandwidth
L-band 1.6/1.5GHz	1.6 GHz	1.5 GHz	Narrowband
C-band 6/4 GHz	5.850-6.425 GHz	3.625-4.200 GHz	500/500 MHz
X-band 8/7 GHz	7.925-8.425 GHz	7.250-7.750 GHz	500/500 MHz
Ku-band 14/12GHz	14-14.5 GHz	11.5-12.75 GHz	500/500 MHz
Ka-band 30/20GHz	27.5-31 GHz	17.7-21.2 GHz	3.5/3.5 GHz

Modulation

The main modulation scheme for the Intelsat/Eutelsat satellite TDMA systems is QPSK. Coherent QPSK uses four phase states providing 2 bits of information for each phase state. Since this is coherent QPSK, the absolute phase states, 0, 90, 180, 270 are used to send the information. The data rate is approximately equal to 120 Mbps, with a bandwidth of 80 MHz. The Intelsat system uses 6/4 GHz, while the Eutelsat uses 14/11 GHz. The RF frequency is downconverted to typical IF frequencies, such as 70 MHz, 140 MHz, and 1 GHz.

To increase the efficiency of the satellite link, various techniques are available. One of these techniques is *digital speech interpolation* (DSI). DSI transmits data signals during the dead times of voice channels or telephone calls. This provides a method of sending data and utilizing the times when the voice is not being sent. Another technique of increasing the efficiency of the link , *digital circuit multiplication equipment* (DCME), combines DSI with a decrease in voice speed from 64 kbps to 32 kbps.

Adaptive Differential Pulse Code Modulation

One of the best and most efficient ways that is used extensively to quantize analog signals is *adaptive differential pulse coded modulation* (ADPCM). The analog signals are sampled at a rate greater than the Nyquist rate. Each of the samples represents a code value for that sample, similar to an A/D. The differential part of the ADPCM describes a process by which the sampled value is compared to the previous sampled value and measures the difference. This difference tells the process if it is greater or smaller value. This difference becomes the code word for that sample. The adaptive part of the modulation scheme means that the step size can be made finer or coarser depending on the last samples. This helps in tracking large analog voltage excursions. If the samples are continually increasing, then the step value is increased. If the samples are continually decreasing, then the step size is decreased (see Figure 10-3).

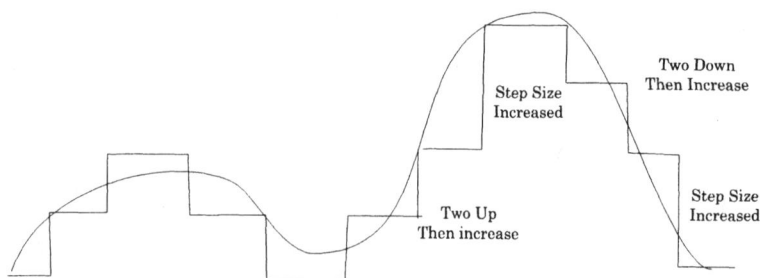

Figure 10-3 ADPCM to convert analog to digital signals.

10.3   Fixed Satellite Service

Satellite communications using geostationary satellites to provide services to the end user is known as *fixed satellite service* (FSS). FSS is a radio communications service between fixed points on the earth using one or more geostationary satellites.

10.4   Geostationary Orbits

Most communication satellites are approximately 22,000 miles above the earth surface. They generally follow a circular orbit on the equatorial plane, circling the earth once every twenty-four hours. By synchronizing the satellite to the earth's rotation, so that the satellite follows the earth at approximately the same speed as the angular rotation of the earth, the satellite will look stationary to a fixed point on the earth (see Figure 10-4).

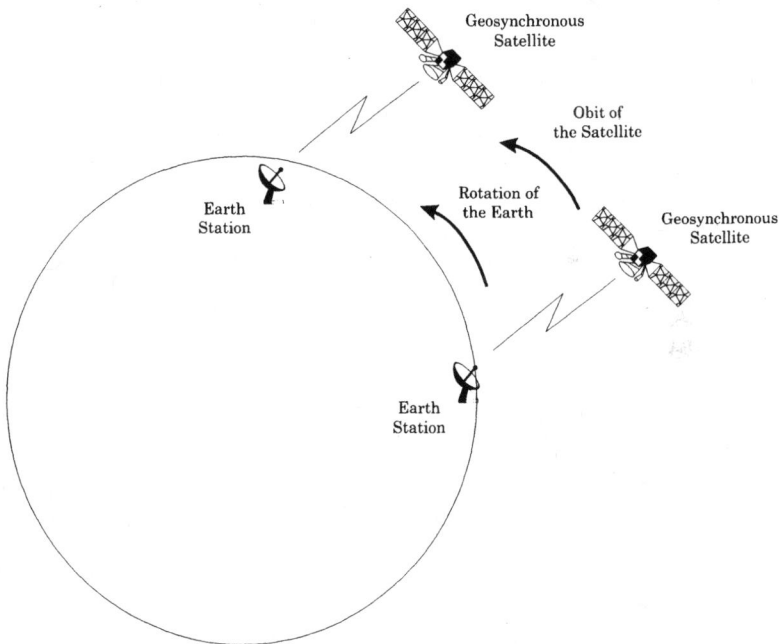

Figure 10-4 *Geostationary satellite orbit.*

This setup provides a continuous link with a given satellite at any time during the day or night. Therefore, the communication link will be available all the time, unless there is a problem with the data link between the earth and the satellite. Another advantage is that the ground equipment becomes less expensive and easier to operate, since tracking a reasonably stationary satellite is much easier than tracking a satellite with high angular velocity.

Although geostationary satellites appear stationary, the satellite actually drifts in a figure 8 pattern north and south of the equator. Satellites are designed to help mitigate this drift, so that the earth stations can be stationary without automatic tracking controls. After these satellites are in orbit for close to space lives, then this figure 8 pattern becomes much larger.

10.5   Ground Station Antennas

Even though the satellite appears to be stationary for a geostationary orbit, large antennas with very narrow beam widths generally require an automatic tracking device to provide the best performance. Smaller antennas generally do not require a tracking device. Digital satellite systems for television and movies use a small antenna and do not require any tracking device. Once the antenna is installed at the site it does not have to be adjusted for several years.

The gain of a ground station antenna is largely dependent on the diameter of the antenna and is also frequency related. The following equation for the gain of the parabolic antenna is often used for satellite communications systems:

$$G_t = 10\log\left[n(\pi(D)/\lambda)^2\right] \qquad 10.1$$

where
   n = efficiency factor < 1,
   D = diameter of the parabolic dish,
   λ = wavelength.

A low-cost system operating in the Ku and Ka bands using small antennas approximately 1 to 2 meters in diameter is the *very small aperture terminal* (VSAT). VSATs provide two-way communications to a central location, the hub. They are used mainly for businesses, schools, and remote areas and connect remote computers and data equipment to the hub via satellites.

Types of Antennas Used in Satellite Communication Systems

Satellite communications uses three types of antenna systems.

1. *Primary focus antenna system.* The feed is positioned in front of the primary reflector, and the signal is reflected once from the feed to the intended direction of radiation. The single reflector antenna system is generally less expensive and provides the simplest design (see Figure 10-5 a).

2. *Cassegrain antenna system.* This antenna system uses a dual reflector arrangement. The feed comes from the back of the primary reflector, sends the signal to a convex subreflector, which is mounted in front of the primary reflector. The signal is fed to the primary reflector where it is reflected in the desired direction of radiation (see Figure 10-5b). Cassegrain antennas are more efficient than primary focus antennas because the subreflector can be adjusted or formed to optimize the signal reflection and to focus less energy towards the blockage of the subreflector, which falls in the path of the primary reflector. Cassegrain antenna systems are also easier to maintain, since the feed horn is located at the base of the reflector, which provides easy access as compared to having the feed horn out on structures in front of the reflector.

3. *Gregorian antenna system.* This antenna system is also a dual reflector system and is very similar to the Cassegrain antenna. The main difference is that the subreflector is concave instead of convex. The feed is mounted on the rear of the primary reflector and sends the signal to the subreflector. The signal is reflected in the concave subreflector that causes a crossover of the reflected signal. The reflected signal is sent to the primary reflector where it is reflected in the desired direction (see Figure 10-5c). Because of this crossover, the subreflector is required to be at a larger distance from the primary reflector.

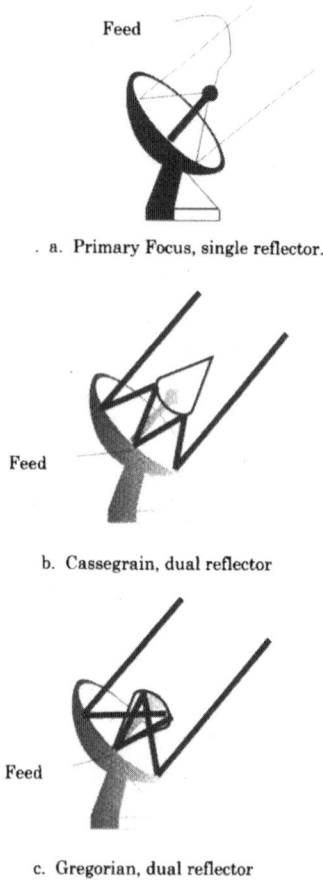

a. Primary Focus, single reflector.

b. Cassegrain, dual reflector

c. Gregorian, dual reflector

Figure 10-5 *Types of antennas for satellite communication systems.*

10.6 Noise and the Low-Noise Amplifier

The LNA is critical to the design of a satellite communications system. The system is improved directly for each improvement in the LNA. The LNA is the main element that sets the noise for the receiver. Unless there are large losses after the LNA, or the bandwidth becomes larger for some reason, which both of these factors may affect the overall noise figure of the system, the LNA is the determining element for the noise figure. Therefore, except for minor adjustments, the LNA sets the noise figure for the receiver. The complete noise factor equation is

$$F_t = F_1 + \left[(F_2 + Losses) - 1 \right] / G_1 + \left[(F_3 + Losses) - 1 \right] / G_1 G_2 \qquad 10.2$$

where

F_t = total noise factor,
F_1 = noise factor of the LNA,
F_2 = noise factor of the next amplifier stage,
F_3 = noise factor of the third amplifier in the receiver,
G_1 = gain of the LNA,
G_2 = gain of the next amplifier after the LNA,
Losses = losses between the different stages.

The noise figure is calculated using the noise factor as follows:

$$NF = 10 \log(FT) \qquad 10.3$$

The noise figure is the LNA (and all additional losses from the following stages according to the Equation 10.2) since the losses in the preceding stages between the antenna and the LNA are included in the budget. If there is loss between the antenna and the LNA, the signal level will be affected. The noise remains the same unless there is a change in temperature from the input of the loss to the output, which is generally not the case. Once the signal passes through the LNA, a bandwidth is established, if not established before, and the signal is amplified well above KTB noise. The noise figure is the noise power that is added to the KTB noise power and is generated by the LNA.

The standard noise equation for calculating the noise after the LNA is

$$N = kT_0 BF \qquad 10.4$$

where

T_0 = nominal temperature (290 degrees K),
F = noise factor,
k = Boltzman's constant,
B = bandwidth.

The noise figure is the noise factor in dB or $10 \log F$. The noise factor is used as a multiplier and the noise figure is in dB, which is additive. The noise will be $kT_0 BF$ plus any effects due to the difference in temperature, with F being the receiver noise factor.

$$n = kT_0 BF + \left(kT_s B - kT_0 B \right) = kT_0 BF + kB \left(T_s - T_0 \right) = kT_0 BF_t \qquad 10.5$$

Solving for F_t,

$$F_t = F + (T_s - T_0) / T_0 \qquad 10.6$$

where
T_s = sky temperature (52 degrees).

If $T_s = T_0$, then Equation 10.2 applies. Since T_s is generally less than T_0, the noise will be less than the standard kT_0BF and the noise factor is reduced by $(T_s - T_0)/T_0$.

Equivalent Temperature Analysis

A satellite system can be evaluated using equivalent noise temperature, which is is another method of establishing link budgets. Most link budgets for terrestrial systems use noise and signal power and the standard KTBF at the output of the LNA, with the losses from the antenna affecting the signal level in the link on a one for one basis. However, for satellite transmission systems, the analysis uses equivalent temperatures and converts the noise factor of the LNA to an equivalent temperature as follows:

$$T_e = (F - 1)T_0 \qquad 10.7$$

where
T_e = equivalent temperature do to the increase in noise of the noise factor of the LNA,
F = noise factor of the LNA (noise figure = 10log(noise factor),
T_0 = 290 degrees K.

The noise factor is determined by solving for F:

$$T_e = (F - 1)T_0 = T_0F - T_0$$
$$F = (T_e + T_0) / T_0 \qquad 10.8$$

Equation 10.7 shows that the noise factor, which is the increase in noise for the receiver, is equal to the additional temperature T_e that is added in front of an ideal receiver that produces the same amount of increased noise. Therefore, if the noise figure of the LNA is, for example, .561 dB, the noise factor is 1.14 and the equivalent temperature T_e is equal to 40 K (using Equation 10.7. That is the noise power at the output of the LNA is equal to kT_0F if the bandwidth and gain is ignored. The input noise power of the LNA is equal to kT_0 with the same assumptions. Therefore, the increase in noise power due to the noise factor of the LNA is equal to

$$\textit{Increase in noise power} = N_i = kT_0F - kT_0 \qquad \text{10.9}$$

Since the equivalent noise increase is desired for the link budget, then the equivalent noise T_e is solved by eliminating k as follows:

$$T_e = T_0F - T_0 = (F-1)T_0 \qquad \text{10.10}$$

The losses in the system between the antenna and the LNA need to be converted into temperature (see Figure 10-6). This is done by taking the difference in noise power due to the total losses as follows:

$$\textit{Noise power difference of attenuator} = N_d = kT_a - kT_a / L_T \qquad \text{10.11}$$

$$\textit{Temperature difference} = T_d = T_a - T_a / L_T = T_a(1 - 1/L_T)$$

where
   T_d = temperature difference due to the losses or attenuation,
   L_T = total losses,
   T_a = temperature of the losses or attenuation, generally = T_0 = 290 K.

Note that if $a = 1$, there are no losses and $T_d = 0$. As a approaches infinity, $T_d = T_a = T_0$. Therefore, the total equivalent temperature out of the antenna for the receiver is:

$$T_r = T_e + T_d = (F-1)T_0 + T_0 - T_0 / L_T = T_0\left[(F-1) + 1 - 1/L_T\right] \qquad \text{10.12}$$

assuming T_0 is the temperature of the losses.

Typical values of T_r range from 70 K for C-bands to greater than 200 K for Ku-bands for small satellite stations. Lower values can be achieved using stabilization.

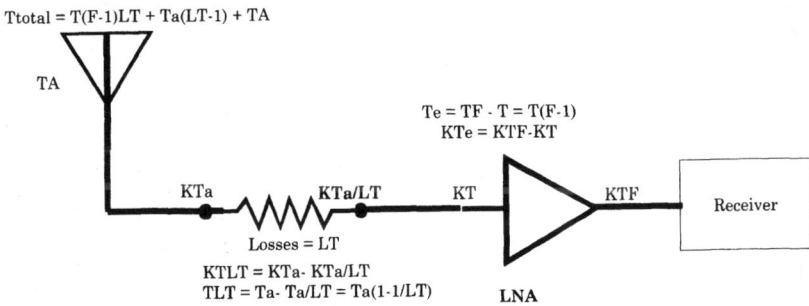

Figure 10-6 *Noise temperature analysis.*

To find the total equivalent temperature for a system, the noise due to the antenna needs to be included. The antenna temperature is determined by integrating the gain and noise temperature in the direction of the antenna.

The noise temperature of both the LNA and the temperature due to the losses are multiplied by the total losses to provide the temperature equivalent at the location of the antenna. Then, the noise of the antenna is then added.

$$T_{total} = L_T(F-1)T_0 + T_a(L_T - 1) + T_a \qquad 10.13$$

This is the total equivalent noise temperature for the system, as shown in Figure 10-5. Once the temperature of the antenna is calculated, a figure of merit, which is used to evaluate different satellite systems, can be used to determine the quality of the receiver. The figure of merit is equal to the gain of the antenna divided by the total temperature at the antenna, G/T. The G/T is usually given as a dB/K. Since the gain of the antenna is usually expressed in dBi, G/T is calculated as follows:

$$G/T = G_d B_i - 10\log(T_{total}) \qquad 10.14$$

A typical value is approximately 30 dB/K, but this value will vary tremendously with the size of antenna, frequency, LNA, losses and other factors.

10.7   The Link Budget

The link budget uses the noise and LNA along with the signal level to determine the range and quality of the satellite link. To calculate the signal or carrier level at the input to the receiver, an effective isotropic radiated power EIRP is determined from the transmitter.

Effective Isotropic Radiated Power
The EIRP out of the transmitting antenna is power output of the power amplifier PA, losses from the PA through the antenna, and the gain of the antenna. The power out of the PA is usually expressed in dBw, 10 log (watts).

Several types of PAs are used for transmitting signals to the satellite. They include *traveling wave tubes* (TWTs), Klystrons, and low-cost *field effect transistor* (FETs) amplifiers. TWTs offer wide bandwidths, up to 500 MHz and higher, good group delay, and the ability to handle many signal inputs. The Klystrons are expensive, high-power only PAs with narrow bandwidths around the 40 to 80 MHz range. The FET *solid state* (SS) amplifiers offer low-cost solutions and include many of the features that the other technologies offer, though generally not for extremely high power outputs.

A general power requirement for different types of signals include 1 watt per channel for telephone signals and 1 kilowatt per television carrier.

The losses are negative (in dB) and are added to the PA output power. The antenna gain is added to the final result (in dB). Some of the losses that could be included in a satellite system are

1. L_{tll} = coaxial or waveguide line losses in dB.

2. L_{comp} = components between the PA and the antenna.

3. L_{tr} = radome losses on transmitter antenna. The radome is the covering over the antenna to protect the antenna from the outside elements that can cause losses.

4. L_{tpol} = polarization loss of antenna. Many antennas are polarized, horizontal, vertical, or circular, right-hand circular and left-hand circular. This defines the spatial position or orientation of the electric and the magnetic fields. A loss is inherent due to polarization and disturbances in the polarization.

5. L_{tfoc} = focusing loss or refractive loss. This is a loss when the antenna receives signals at low elevation angles.

6. L_{tpoint} = mispointed loss. This is caused by transmitting and receiving directional antennas that are not exactly lined up. Therefore, the gains of the antennas do not add up without a loss of signal power.

The total losses included in the link budget are

$$L_{ta} = L_{tll} + L_{comp} + L_{tr} + L_{tpol} + L_{tfoc} + L_{tpoint} \qquad 10.15$$

The antenna has gain, and if a parabolic dish antenna is used, then the gain is calculated by using Equation 10.1. This gain is expressed in dB so that it can be added to achieve the effective isotropic radiated power EIRP. The EIRP is calculated as follows:

$$EIRP = P_t + L_{ta} + G_{ta} \qquad 10.16$$

where
P_t = power output of PA in dB,
L_{ta} = total attenuation in the transmitter (negative).
G_{ta} = gain of the transmitter antenna.

Propagating Channel

The propagating channel is the path of the RF signal that is transmitted from the transmitter antenna and received by the receiver antenna, i.e., the signal in space that is attenuated by the channel medium. The main contributor to the channel loss is *free-space attenuation*. Other factors, such as the propagation losses and multipath losses are fairly small compared to freespace loss.

As a wave propagates through space, loss occurs due to dispersion, i.e., the "spreading out" of the beam of radio energy as it propagates through space. This loss is consistent relative to wavelength, which means that it increases with frequency, as the wavelength becomes shorter. It is called *free-space loss* or *path loss* and is related to both the frequency and the slant range, the distance between the transmitter and receiver antennas. The equation is

$$A_{fs} = 20\log\left[\lambda / \left(4\pi R\right)\right] = 20\log\left[c / \left(4\pi R_f\right)\right]$$ 10.17

where
  λ = wavelength,
  R = slant range,
  c = speed of light, 300×106 meters/sec (R is in meters),
  f = frequency.

The free space loss for satellite system operating at X-band is approximately 200 dB. There are other losses, depending on the conditions in the atmosphere, such as clouds, rain and humidity that need to be included in the link budget. The three main losses are:

1. Cloud loss — loss due to water vapor in clouds.
2. Rain loss — due to rain, dependent on operating region and frequency.
3. Atmospheric absorption (oxygen loss) — loss due to the atmosphere.

These values are usually obtained from curves and vary from day to day and from region to region. Each application is dependent on the location and a nominal loss, generally not worst case, is used for the link analysis.

Multipath Losses

Whenever a signal is sent out in space, it can either travel on a direct path or it can take multiple indirect paths. The most direct path the signal can take is with the least amount of attenuation.

The problem with multipath is that the signal takes both paths and interferes with each other at the receiving end. The reflected path has a reflection coefficient that can change the phase and amplitude, and the path length gives the reflected signal a different phase. If the phase of the reflected path is different, say 180 degrees out of phase, from the direct path, and the amplitudes are the same, the signal is cancelled out and the receiver sees very little to no signal. Most of the time the reflected path is attenuated, depending on the reflection coefficient and the type of multipath. It does not completely cancel out the signal but can alter the amplitude and phase of the direct signal. Therefore, multipath can affect coverage and accuracy where reliable amplitude or phase measurements are required. The losses are included in the link budget as follows:

L_{multi} = Losses due to multipath cancellation of direct path signal in dB.

This loss is generally hard to quantize and usually associated with a probability number, such as two sigma. The multipath is constantly changing and certain conditions can adversely affect the coverage and the phase measurement accuracy.

Receiver Antenna Losses

There are antenna losses for the receiver that are very similar to those for the transmitter. Some of the commonly occurring losses are

1. L_{rll} = coaxial or waveguide line losses in dB.

2. L_{rcomp} = components between the PA and the antenna.

3. L_{rr} = radome losses on transmitter antenna. The radome is the covering over the antenna to protect the antenna from the outside elements that can cause losses.

4. L_{rpol} = polarization loss of antenna. Many antennas are polarized, horizontal, vertical, or circular, right-hand circular and left-hand circular. This defines the spatial position or orientation of the electric and the magnetic fields. A loss is inherent due to polarization and disturbances in the polarization.

5. L_{rfoc} = focusing loss or refractive loss. This is a loss when the antenna receives signals at low elevation angles.

6. L_{rpoint} = mispointed loss. This is caused by transmitting and receiving directional antennas that are not exactly lined up. Therefore, the gains of the antennas do not add up without a loss of signal power.

The total losses for the antenna can be calculated by adding all of the losses together (assuming that their values are in dB).

$$L_{ra} = L_{rll} + L_{rcomp} + L_{rr} + L_{rpol} + L_{rfoc} + L_{rpo\,int} \qquad 10.18$$

Received Signal Power at the Input to the LNA

The received signal level at the input to the LNA is calculated as follows:

$$P_s = EIRP + A_{fs} + L_p + L_{ra} + G_r + L_{multi} \qquad 10.19$$

This equation makes the assumption that all the losses are negative and the gain and EIRP are positive. Also, the above equation assumes that all the parameters are in either dB, dBm or dBw.

Other losses included in the link budget are the phase noise or jitter of all of the oscillator sources in the system, including the up and down conversion oscillators, the code and carrier tracking oscillators and match filter and A/D oscillators. Another source of implementation loss is the detector process including nonideal components and quantization errors. This loss directly affects the receiver's performance and degrades the link budget.

L_i = implementation loss (departure from ideal, detector implementation, phase tracking of phase lock loops).

Table 10-2 shows a typical link budget for satellite communications. It evaluates the tradeoffs of the system to ensure that the receivers on either end have enough S/N for reliable communications.

Carrier Power/Equivalent Temperature

The power received by the receiver is called the *carrier power* (C). This carrier power is compared to an *equivalent temperature* (C/T). Combining the losses, dividing both sides by temperature, and converting all values to actual values, a simplified form of Equation 10.19 is

$$C/T = EIRP \quad G_r/(L_T\ T) = G_r/T \quad EIRP/L_T \qquad 10.20$$

where
C/T = carrier power/equivalent temperature,
G_r/T = figure of merit,
$EIRP$ = effective isotropic radiated power,
LT = total losses from EIRP of the transmitter to the receiver.

Table 10-2 *Link budget for the uplink from the earth station to the satellite.*

Link Budget Analysis:				
	Slant Rng(Km)	**Freq.(GHz)**	**Power(W)**	
Enter Constants.........	35000	6	10	
Enter Inputs..............	**Inputs**	**Power Levels**		**Temp**
Transmitter	Gain/Loss (dB)	Sig.(dBm)	Noise(dBm)	Kelvin
Tran.Pwr(dBm)=		40		
Trans. line loss =	-0.5	40		
Other(switches)	-0.5	39		
Trans Ant Gain =	47	86		
Ant. Losses=*	-1	85		
EIRP		85		
Channel				
Free Space Loss =	-198.89	-113.89		
Rain Loss =	-0.20	-114.09		
Cloud Loss =	-0.10	-114.19		
Atm loss(etc) =	-0.10	-114.29		
Multipath Loss=	-2.00	-116.29		
Receiver				
Rx Ant Gain =	47.00	-69.29		
G/T dB/K				**25.20**
Total Noise Temp at Ant.				152
Antenna Noise TA				30
Ant. losses =	-0.02	-69.31		
Other(switches)	-0.25	-69.56		
Rec. Line loss =	-0.25	-69.81		
Total Losses LT	-0.52			
RF BW(MHz)**	100.00		-94.00	
Total Noise Temp. at Rec.				134
Equiv. Temp. Te =				75
LNA Noise Fig. =	1.00		-93.00	
LNA Gain =	25.00	-44.81	-68	
LNA levels =		-44.81	-68.00	
Receiver Gain =	60.00	15.19	-8.00	
Imp. Loss	-4.00	11.19		
Detector BW	10.00		-18	
Det. Levels =		11.19	-18	
S/N =		29.19		
Req. Eb/No				
Req. Eb/No			12	Pe=10exp-8
Coding Gain =	4.00	8.00		
Eb/No Margin =		21.19		

Power flux density determines the amount of power radiated by the antenna (in one direction) at a large distance per unit of surface area. For an isotropic radiator, the equation equals

$$PDF = P / \left[4(\pi)d^2 \right]$$

10.21

where

PDF = power flux density,

P = output power,

d = distance.

When the gain is added, we obtain the PFD for a gain antenna. With the satellites increased power and more sensitive receivers, the earth stations are becoming less costly and smaller. For example, the Intelsat V satellite contains 50 transponders on board and operates with just 5 to 10 watts of power. The total bandwidth for the Intelsat V is 500 MHz to provide high-speed data for multiple users.

10.8 Multiple Channels in the Same Frequency Band

In order to utilize the band more efficiently and provide more data or users, two different schemes are employed. The first scheme uses beam pointing at two different points on earth using the same satellite. Each beam is focused on one area of the earth so that the same antenna can be used for two systems in the same band and frequency. The narrower the beamwidth, the less coverage area is illuminated on the earth surface. However, this provides for more EIRP so that the earth stations can be less expensive.

Another way to increase data capacity for a given band is to use polarization. If orthogonal polarizations are used for two different channels, the channels can use the same antenna and bandwidth with minimal interference to each other. In theory, horizontal and vertical polarizations are orthogonal, and LHCP and RHCP are also orthogonal. Cross-polarization degrades the separation of the two channels that are orthogonally polarized. Frequently, orthogonal polarization is used to provide increased isolation between the transmitter and the receiver for transceiver operation.

In order for a system to use polarization for frequency re-use in a satellite communication system, the isolation between the channels should be at least 25 to 30 dB. The main causes that degrade this isolation are the Faraday effect (earth's magnetic field) and atmospheric effects (rain or ice crystals). Also multipath can alter the polarization during reflection of the signals on a surface and propagation through the troposphere or ionosphere can cause disturbances in the polarizations.

10.9 Multiple Access Schemes

Multiple access allows multiple users in the same bandwidth, satellite, and antenna system. Two basic methods are used to allow for multiple users.

The first method is a multiplexing scheme called *preassigned multiple access* (PAMA). This method permanently assigns a user to a channel or time. An example of a PAMA is using time slots and assigning a given time slot to each end user. Since they are permanently assigned, the users will have that given time slot and will be multiplexed with all of the other users similar to a TDM used in other communication links.

The second method of multiplexing is *demand assigned multiple access* (DAMA). This system is a true multiple access scheme similar to TDMA used in other communication links. It is used on an "as needed basis," i.e., each user takes any time slot when needed.

Another method of providing multiple users is FDM/FDMA. These methods use frequency division to provide multiple users for one band. In systems that incorporate FDMA, the channels are separated using different frequency slots for different users.

10.10 Propagation Delay

One of the problems with real-time communications using satellites is due to the large distance between the earth station and the satellite. The approximate propagation delay equals 275 ms; the two-way propagation delay is double that time (550 ms). Since this propagation delay is so long, echo cancellation is vital to the quality of the communication link especially for voice applications. Video applications, in particular one way broadcasting, generally does not need echo cancellation techniques.

10.11 Cost

The cost of using satellites is determined by the type of transmission, type of signals sent, and the length of time of transmission. Transmission types include preassigned, demand assigned, or occasional. The preassigned transmission type is the most costly, since it ties up an entire multiple access slot for all of the time. The occasional type is the least costly because it is only billed for the time that it is used.

Voice, video, or data are sent using the satellite link. Many systems, the Internet, for example, do not include voice as an option. For more information on the cost of use of the satellites and detailed charges refer to the Intelsat tariff handbook.

10.12 Regulations

ITU radio regulation provides the rules for satellite communications to avoid interference and confusion. The main specifications supplied by the ITU are frequency band allocations, power output limitations from either the earth station or the satellite, minimum angles of elevation for earth station operation and pointing accuracy of antennas.

10.13 Types of Satellites Used for Communications

The Inmarsat satellite system is mainly used for maritime communications for ships and shore earth stations. Currently, it is being used for broadband communications and extended coverage. The satellites cover the main bodies of water, but earth stations do not have to be located on the shore, as long as the satellites are visible. The Inmarsat satellites operate on C-band and L-band. The FSS gateway for telephone service is via the Inmarsat satellites.

The Intelsat satellite provides broadband communications, including digital data, video, telephones and other communication methods.

With the push towards providing broadband communications many companies, such as Astrolink, CyberStar, Skybridge, Spaceway, and Teledesic are working to develop both Ku and Ka band satellite communication systems. Hybrids satellite constellations are used to provide both Ka and Ku band operation with data capacities ranging from 30 Gbps to systems providing much greater data capacities.

10.14 System Design for Satellite Communications

The main design criteria for the satellite are the transponder bandwidth, EIRP, and G/T, gain to noise temperature level (gain of the receiving antenna in the direction of the received signal vs. the receiving system noise temperature). Typical figure of merit values for 4 GHz range from 41 dB/K using a 30-meter antenna with a parametric amplifier to 23 dB/K for a 4.5-meter antenna using an FET amplifier. For a space station receiver at 6 GHz, a typical figure of merit ranges from 19 dB/K, using a FET transistor LNA and providing a wide coverage area using a wide beamwidth, to –3 dB/K for a pencil beam antenna. For the earth station, the main design criteria are G/T, antenna gain, system noise temperature, and transmitted power. The overall system design criteria are operation frequency bands, modulation methods, multiple access parameters, system costs, channel capacity, and overall system performance.

10.15 Summary

Satellites are becoming a solution to broadband communications and connecting the "last mile." The satellite connection consists of a remote earth station, a satellite, and another earth station. This triangle forms a two-way communications link to provide the remote earth station access to the Internet, video, voice and data at high data rate.

Four different bands are used by the satellite systems and ground systems: C, X, Ku and Ka. The latter is becoming popular for commercial use in the broadband arena. The geostationary orbit is used so that the ground station tracks a fairly stationary transceiver and the satellite appears to be stationary.

A link budget determines the power, the gains and losses in the system as well as the figure of merit, G/T. Multiple access schemes are used to allow multiple users on the same band. Costs are associated with the type of system and the length of use.

10.16 References

1. Bullock, Scott R. *Transceiver System Design for Digital Communications*. Atlanta: Noble Publishing, 1995.

2. "Satellite Communications Handbook, Fixed Satellite Service." *International Telecommunications Union, International Radio Consultative Committee,* Geneva (1988).

3. Inglis, Andrew F. *Electronic Communications Handbook.* New York: McGraw-Hill, 1988.

4. Crane, Robert K. "IEEE Transactions on Communications." vol. Com-28, no. 9 (September 1980).

5. Schwartz, Mischa. *Information, Transmission, Modulation and Noise.* New York: McGraw-Hill, 1980.

6. Haykin, Simon. *Communications Systems*. New York: John Wiley & Sons, 1983.

7. Holmes, Jack K. *Coherent Spread Spectrum Systems*. New York: John Wiley & Sons, 1982.

8. Bullock, Scott R. "Use Geometry to Analyze Multipath Signals." *Microwaves & RF* (July 1993).

9. Bao-yen Tsui, James. *Microwave Receivers with Electronic Warfare Applications.* New York: John Wiley & Sons, 1986.

Index